Plants of Central Asia

Volume 8b

Plants of Central Asia

Plant Collections from China and Mongolia

Volume 8b

Legumes
Genus: Oxytropis

V.I. Grubov

CRC Press
Taylor & Francis Group
Boca Raton London New York

CRC Press is an imprint of the
Taylor & Francis Group, an **informa** business

A SCIENCE PUBLISHERS BOOK

ACADEMIA SCIENTIARUM URSS
INSTITUTUM BOTANICUM nomine V.L. KOMAROVII
PLANTAE ASIAE CENTRALIS
(secus materies Instituti botanici nomine V.L. Komarovii)
Fasciculus 8b
OXYTROPIS DC.
Confecit : V.I. Grubov

First published 2003 by Science Publishers Inc.

Published 2019 by CRC Press
Taylor & Francis Group
6000 Broken Sound Parkway NW, Suite 300
Boca Raton, FL 33487-2742

© 2003, Copyright reserved
CRC Press is an imprint of Taylor & Francis Group, an Informa business

First issued in paperback 2019

No claim to original U.S. Government works

ISBN 13: 978-0-367-44687-1 (pbk)
ISBN 13: 978-1-57808-120-2 (hbk)
ISBN 13: 978-1-57808-062-5 (Set)

**Visit the Taylor & Francis Web site at
http://www.taylorandfrancis.com**

**and the CRC Press Web site at
http://www.crcpress.com**

Library of Congress Cataloging-in-Publication Data
Rasteniia TSentral'noǐ Azii. English
  Plants of Central Asia: plant collections from China
  and Mongolia
  /[editor-in-chief. V.I. Grubov].
    p. cm.
  Research based on the collections of the V.L.
  Komarov Botanical Institute.
  Includes bibliographical references.
  Contents: V.8b. Legumes. Genus: Oxytropis
  ISBN 1-57808-120-3 (v.8b)
  1. Botany-Asia, Central. I. Grubov, V.I. II.
Botanicheskiǐ institut im. V.L. Komarova. III. Title.
QK374, R23613 2002
581.958-dc21   99-36729
                                              CIP

Translation of: Rasteniya Central'nov Asii, vol. 8b, 1989;
                   JPCPH Press, Mir i Sem'ya

ANNOTATION

This volume, 8b, of the illustrated lists of vascular plants of Central Asia (within the People's Republics of China and Mongolia) is entirely devoted to the treatment of crazyweeds (genus *Oxytropis*), one of the two largest in the family Leguminosae (153 species). A key to the species precedes their listing.

The preparation of camera ready copy of the Russian edition was sponsored by Russian Foundation for Basic Research (Grant 93-04-20637).

Russian edition printed with the support from St.-Petersburg State Chemical-Pharmaceutical Academy.

V.I. Grubov
Editor-in-Chief
Reviewers
T.V. Egorova and N.N. Tzvelev
St.-Petersburg
1998

PREFACE

This volume is exclusively devoted to one of the largest genera of Leguminosae and the largest in the Central Asian flora—genus *Oxytropis* DC. Within the territory covered by the present series, the genus is represented by 153 species, out of the 175 species in Central Asia as a whole, (22 species are found only in the Central Asian part of the former USSR) i.e., more than one-half of all known species of the genus. This genus is extremely diverse and rich in endemics: of the 153 species covered in this book, 62 are endemic to the region and 78, or 44.3% of the total number of species, endemic to the Central Asian territory. This signifies that Central Asia, along with Fore Asia, is the most important centre of the speciation of genus *Oxytropis*. It includes all the 6 subgenera and most of the sections, i.e., 16. In terms of the number of species, subgenus *Eumorpha* (Bge.) Abdus. (= *Euoxytropis* Bge.) with 91 species and subgenus *Oxytropis* (= *Phacoxytropis* Bge.) with 51 species occupy the prominent position while the largest sections are *Baicalia* (24 species), *Xerobia* (20) and *Orobia* (18) in the former subgenus and *Ianthina* (21) and *Mesogaea* (18) in the latter.

The largest number of species, including endemics, is concentrated in the mountain systems of Mongolian Altay, Tien Shan, Nanshan and Himalayas. Further, mesophylic alpine-meadow species predominate in subgenus *Oxytropis*: in section *Oxytropis*, these are mainly Himalayan-Tibetan (*O. biflora, O. glacialis, O. proboscidea, O. sericopetala, O. tatarica*) or even Tibetan-Tien Shan species (*O. globiflora, O. pauciflora, O. platysema*) and in section *Ianthina*, the prominent species are Tien Shan (*O. chantengriensis, O. kumbelica, O. larionovii, O. penduliflora, O. rupifraga*), Ti-betan-Himalayan-Tien Shan (*O. densa, O. lutchensis*), Pamiro-Alay-Tien Shan (*O. melanotricha, O. merkensis*) and Altay species (*O. krylovii, O. ladyginii, O. saposhnikovii*). The picture is similar in section *Mesogaea* but it contains as well species with a wide distribution range—Eurasian (*O. lapponica, O. pilosa*), north-Asian (*O. glabra*) and even Asian-North American (*O. deflexa*). The series of north Tibetan-western Chinese species (*O. gueldenstaedtioides, O. kansuensis, O. melanocalyx, O. ochrocephala*) is of particular interest in this context. Apart from well-recognised floristic contacts between Altay, Junggar Ala Tau and Tien Shan, Tien Shan and Pamir, Tien Shan and Himalayas which confirm the geographic spread of genus *Oxytropis*, Tien Shan-Nanshan contacts also are detected here. Floristic relation among mountain systems of West China through Nanshan with Eastern and Northern Tien Shan are at once clearly visible in section *Mesogaea*. Thus, *O. kansuensis* bears close affinity with Tien Shan *O. meinshausenii* and with western Chinese *O. gueldenstaedtioides* and *O. ochrocephala* to form a natural series. *O. hirsutiuscula* has Pamir-eastern Tien Shan-Nanshan as its distribution range. These relations demonstrate the close genetic affinity of *O. imbricata* (Nanshan) and *O. merkensis* (Tien Shan) of section *Ianthina*.

In subgenus *Eumorpha*, sections *Orobia*, *Ortholoma* and *Xerobia* are represented mainly by montane-steppe species with distribution ranges in Altay, Junggar-Tarbagatai, Tien Shan and Mongolia. Endemic Mongolian species are particularly conspicuous in section *Xerobia* (*O. diversifolia*, *O. junatovii*, *O. klementzii*, *O. micrantha*, *O. monophylla*, *O. potaninii*, *O. rhizantha*). The largest section of this subgenus, *Baicalia*, is most characteristic of Mongolia. It comprises predominantly montane-steppe and steppe Siberian-Mongolian species (*O. lanata*, *O. lanuginosa*, *O. lasiopoda*, *O. myriophylla*, *O. oxyphylla*, *O. prostrata*, *O. selengensis*, *O. viridiflava*) as well as strictly regional Mongolian endemics (*O. mongolica*, *O. pavlovii*, *O. pumila*, *O. ramosissima*, *O. sutaica*). This section, however, also contains species with a more extensive distribution range: northern China-Mongolia (*O. bicolor*, *O. ochrantha*), Kazakhstan (*O. rhinchopysa*), Altay-Saur-Mongolia (*O. oligantha*), eastern Tien Shan-Mongolian Altay (*O. heterophylla*), Junggar-Tien Shan (*O. chionobia*) and narrow endemics eastern Tien Shan *O. przewalskii*, Junggar-Tarbagatai *O. fetissovii* and Saur *O. saurica*.

Species of oligotypic section *Polyadena* inhabit most severe regions: alpine deserts and desert steppes. These are Tibetan *O. chiliophylla* and *O. falcata* and Mongolian *O. microphylla* and *O. trichophysa*. Moreover, the distribution ranges of *O. falcata* and *O. microphylla* are disjunced: while the former has an isolated enclave in north-western Mongolian Altay, three independent ranges are known for the latter: in Mongolian Altay, Eastern Gobi and southern Dauria.

Four other subgenera of *Oxytropis* are poorly represented. Subgenus *Physoxytropis* is represented by a single species in Central Asia: endemic Mongolian montane-steppe *O. bungei* distributed in Mongolian and Gobi Altay and along the southern slope of main Hangay mountain range and monotypic subgenus *Triticaria* represented by meadow-steppe Manchurian-northern Chinese *O. hirta* entering Eastern Mongolia.

Fore Asian subgenus *Ptiloxytropis* was unexpectedly represented by endemic desert *O. sacciformis* that has only recently been detected in Eastern Gobi. Two other species of the subgenus—*O. bella* and *O. trichosphaera*—only enter Pamir in Central Asia from the west.

Finally, *Traganthoxytropis*, evidently an artificial subgenus of spiny-shrubby crazyweeds, petrophytes and psammophytes, includes ortuz *O. aciphylla* which is conspicuous gregarious. This is a lone coenosis-forming *Oxytropis* and forms small mats or cushions in deserts over large expanses of sandy and sandy-rubbly Gobi plains and mountain slopes. Its range extends form Zaisan lake to Ordos, Qaidam and Qinghai through Junggar and Mongolia. Its derivative, the tiny steppe petrophyte *O. kossinskyi*, is a narrow endemic species of central Mongolia. Another extensively distributed species, *O. tragacanthoides*, is found on cliffs with its range extending from Altay and Sayan to Eastern Tien Shan and northern Tibet. The remaining two species of this subgenus, Mongolian-Altay *O. acanthacea* and Junggar-Tarbagatai *O. hystrix*, inhabitants of rocks and rocky slopes, are strict endemics.

Crazyweeds have no great economic importance and some are of minor significance as fodder plants (*O. aciphylla, O. myriophylla, O. filiformis*) while the widely-distributed meadow-solonchak *O. glabra* is a favourite of domestic animals but is toxic to all types of cattle causing intense poisoning.

––––––––

Artist S.S. Loseva did the drawings in the Plates for this volume while the distribution ranges were plotted by the author.

O.I. Starikova translated the Chinese references and herbarium labels.

CONTENTS

ANNOTATION   v

PREFACE   vii

TAXONOMY   1
   Special Abbreviations   1
   Genus *Oxytropis* DC.   3

PLATES I TO IV   97

MAPS 1 TO 4   101

INDEX OF LATIN NAMES OF PLANTS   105

INDEX OF PLANT DISTRIBUTION RANGES   109

INDEX OF PLANT DRAWINGS   111

TAXONOMY

SPECIAL ABBREVIATIONS

Abbreviations of Names of Collectors

Bar.	— V.I. Baranov
Chaff.	— J. Chaffanjon
Ching	— R.C. Ching
Chu	— C.N. Chu
Czet.	— S.S. Czetyrkin
Divn.	— D.A. Divnogorskaya
Fedtsch.	— B.A. Fedtschenko
Fet.	— A.M. Fetisov
Glag.	— S.A. Glagolev
Golubk.	— N.S. Golubkova
Gr.-Grzh	— G.E. Grum-Grzhimailo
Grombch.	— B.L. Grombchevski
Grub.	— V.I. Grubov
Gub.	— I.A. Gubanov
Gus.	— V.A. Gusev
Ik.-Gal.	— N.P. Ikonnikov-Galitzkij
Isach.	— E.A. Isachenko (also known as E.A. Volkova)
Ivan.	— A.F. Ivanov
Kal.	— A.V. Kalinina
Kam.	— R.V. Kamelin
Karam.	— Z.V. Karamysheva
Klem.	— E.N. Klements
Krasch.	— I.M. Krascheninnikov
Krasn.	— A.N. Krasnov
Kuan	— K.C. Kuan
Lad.	— V.F. Ladygin
Ladyzh.	— M.V. Ladyzhensky (Ladyzhinsky elsewhere)
Lavr.	— E.M. Lavrenko
Lipsk.	— V.I. Lipsky
Lis.	— V.I. Lisovsky
Litw.	— D.N. Litwinow

Lom.	— A.M. Lomonossov
Merzb.	— G. Merzbacher
Mois.	— V.S. Moiseenko
Nikul.	— R.I. Nikulina
Pal.	— I.V. Palibin
Petr.	— M.P. Petrov
Pob.	— E.G. Pobedimova
Pop.	— M.G. Popov
Pot.	— G.N. Potanin
Przew.	— N.M. Przewalsky
Rachk.	— E.I. Rachkovskaya
Reg. A.	— A. Regel
Rob.	— V.I. Roborowsky
Saposhn.	— V.V. Saposhnikov
Schischk.	— B.K. Schischkin
Serp.	— V.M. Serpukhov
Sold.	— V.V. Soldatov
Sumer.	— I.Yu. Sumerina
Tug.	— A.Ya. Tugarinov
Ulzij.	— N. Ulzijkhutag
Vinogr.	— V.M. Vinogradova
Volk.	— E.A. Volkova (also known as E.A. Isachenko)
Wang	— K.S. Wang
Yun.	— A.A. Yunatov
Zam.	— B.M. Zamatkinov

Abbreviated Names of Herbaria

A	— Arnold Arboretum of Harvard University, Cambridge, USA
AA	— Herbarium of the Botanical Institute of the Academy of Sciences of Kazakhstan, Alma-Ata
B	— Botanisches Museum, Berlin-Dahlem
BM	— British Museum of Natural History, London
BP	— Botanical Department of the Hungarian History Museum, Budapest
BRNU	— Institute of Plant Biology and Herbarium, University, Brno
C	— Botanical Museum and Herbarium, Copenhagen
G	— Conservatoire et Jardin botaniques, Geneve (Geneva)
GB	— Herbarium Goteborg University, Goteborg, Sweden

6

HIMC	— Herbarium, Department of Biology, University of Inner Mongolia, Huhhot [Khuk-khoto(NMU)]
K	— The Herbarium, Royal Botanic Gardens, Kew, Richmond, Surrey, London
LE	— V.L. Komarov Botanical institute, Saint Petersburg (formerly Leningrad), Russia
LINN	— Herbarium, The Linnean Society of London, London
LZDI	— Herbarium, Institute of Desert Research, Academia Sinica, Gansu, Lanzhou
MW	— Herbarium of the Moscow State University, Moscow
NMU	— Khuk-khoto (Acronym not listed in Index Herbariorum)
NWBI	— Herbarium, North-Western Plateau Institute of Biology, Academia Sinica, Shaanxi, Wugung
P	— Museum National d'Histoire Naturelle, Laboratoire de Phanerogamie, Paris
TAK	— Tashkent (Acronym not listed in Index Herbariorum).
PE	— Herbarium, Institute of Botany, Academia Sinica, Beijing
TI	— Botanical Institute, University of Tokyo, Tokyo
TK	— Krylov Herbarium of the State University, Tomsk
UBA	— Herbarium, Institute of Botany, Mongolian Academy of Sciences, Ulan-Bator
US	— United States National Herbarium, Department of Botany, Smithsonian Institute, Washington
W	— Naturhistorisches Museum, Botanische Abteilung, Wien (Vienna)

33. Oxytropis DC.
Astragal. (1802) 53.

1. Small cushion forming prickly shrubs with dense spines formed of lignified rachis (subgenus VI. **Traganthoxytropis** Vass.) ....2.
+ Herbaceous perennial shrubs devoid of spines but, not infrequently, with stems lignified at base or surface shoots ......6.
2. Leaves paripinnate as terminal leaflet is caducous; leaflets thorny, with spinule at tip. Pods hard, nutlet-shaped (section 1. **Lycotricha** Bge.) ....3.
+ Leaves imparipinnate or with whorled leaflets; leaflets not spiny. Pods vesicular..4.
3. Peduncles with 1–2 (3) flowers. Pod 12–15 mm long, exceeding calyx. Spines rigid, not brittle ....149. **O. aciphylla** Ledeb.
+ Peduncle 1-flowered. Pod small, not longer than 10 mm, completely enclosed in indehiscent calyx. Spines slender, brittle ...
.... 150. **O. kossinskyi** B. Fedtsch. et Basil.

4. Leaves with leaflets in whorls of 3–6 each. Pods oblong-ovate, 15–18 mm long. Racemes few-flowered (section 3. **Acanthos** Ulzij.) .. .. 153. **O. acanthacea** Jurtz.

+   Leaves imparipinnate. Pods globose-ovate or subglobose, 17–25 mm long. Racemes 3–4-flowered (section 2. **Hystrix** Bge.) .... 5.

5. Leaflets 3–6 pairs, oblong, 8–12 mm long and 3–5 mm broad, with appressed silvery hairs. Inflorescence umbellate, 3–4-flowered. Pod white-hariy .... 152. **O. tragacanthoides** Fisch.

+   Leaflets 8–11 pairs, linear, 7–8 mm long, 1–1.5 mm broad, ash-grey due to appressed hairs. Inflorescence 2-flowered. Pods subglabrous .... 151. **O. hystrix** Schrenk.

6. Mature pods exserted from and usually rupturing calyx .... 7.

+   Mature pods enclosed in calyx, remain unchanged or inflated. Plant acauline ....151.

7. Pods unilocular, without walls on both sutures, pubescent with simple hairs (subgenus I. **Oxytropis**) .... 8.

+   Pods bilocular, with walls on both sutures, or semibilocular with wall on ventral suture, pubescent with simple or glandular hairs (subgenus II. **Eumorpha** (Bge.) Abdus. .... 60.

8. Stipules more or less strongly adnate to petiole (section 2. **Ianthina** Bge.) .... 9.

+   Stipules not adnate to petiole or barely adnate only at base ... 31.

9. Flowers and pods pendent; peduncles thickened, villous; racemes oblong or oval, highly elongated. Leaflets up to 20–30 pairs .... 10.

+   Flowers not pendent; peduncles not thickened and not villous. Leaflets up to 20 pairs. .... 11.

10. Corolla yellow, 15–20 mm long. Pods with dense silky hairs .... 29. **O. nutans** Bge.

+   Corolla blue, 14–15 mm long. Pods patently hairy .... 30. **O. penduliflora** Gontsch.

11. Corolla 13–23 mm long .... 12.

+   Corolla 5–12 mm long .... 14.

12. Corolla 18–23 mm long. Plant green, pubescent with long, soft patent hairs. Pods erect. .... 26. **O. lutchensis** Franch.

+   Corolla 14–20 mm long. Plant ash-grey or sericeous due to dense pubescence. .... 13.

13. Plant ash-grey due to dense appressed pubescence. Inflorescence 3–5-flowered; calyx 10–12 mm long, with villous black hairs. Pods pendent. ....27. **O. melanotricha** Bge.

+   Plant villous with white hairs. Inflorescence 5–7-flowered; calyx 8–10 mm long, white hairy. Pods erect ....32. **O. rupifraga** Bge.

14. Leaves green; leaflets with scattered long hairs or subglabrous above .... 15.

+ Leaves and entire plant greyish-green, ash-grey or silvery due to dense pubescence; leaflets densely hairy above .... 18.

15. Plant acauline, very small, 1–5 cm tall, subglabrous. Stipules glabrous, lustrous. Inflorescence 2–5-flowered. Corolla 5–7 mm long .... 16.

+ Plant with short (1–4 mm long) procumbent stems, 5–10 to 20 cm tall, with scattered hairs, stipules ciliate along margin. Inflorescence few-flowered. Corolla 9–12 mm long .... 17.

16. Leaflets 5–9 pairs, closely set, oblong, 3–5 mm long. Peduncles shorter than leaves. Standard of corolla orbicular. Pods erect ......
.... 13. **O. brevipedunculata** P.C. Li.

+ Leaflets 3–6 pairs, distant, linear-lanceolate, 5–10 mm long. Peduncles longer than leaves. Standard oblong. Pods pendent or strongly declinate. .... 31. **O. pusilla** Bge.

17. Leaflets 6–7 pairs. Corolla 8–10 mm long. Violet-coloured ....
.... 33. **O. saposhnikovii** Kryl.

+ Leaflets 8–12 pairs. Corolla 9–12 mm long, bright-lilac ....
.... 21. **O. krylovii** Schipcz.

18. Plants small, 2–10 cm tall, densely pubescent plant forming dense flat or cushion-shaped mats, with highly branched surface shoots, frequently covered with sheath of dead stipules and petioles. .... 19.

+ Plants very large (7–15 cm tall), peduncles up to 30 cm tall; dense mats not formed .... 22.

19. Entire plant covered with long patent hairs; leaflets densely lanate. Peduncles scarcely exceeding leaves; inflorescence densely-capitate. Pods erect .... 20.

+ Plant ash-grey, densely covered with white appressed hairs; leaflets canescent. Peduncles longer than leaves. Pods pendent ..
.... 21.

20. Plant with villous grey-hairs forming cushion-shaped mats; dead stipules closely overlapping covering surface shoots. Pods oblong-lanceolate, 12–15 mm long, trigonous, radially divaricate. Corolla about 6 mm long. Leaflets lanceolate, 2–3 (5) mm long, 4–6 pairs .... 16. **O. densa** Benth.

+ Plant villous-hirsute, 6–10 cm tall. Leaflets oblong, 7–10 mm long and 3–4 mm broad, 7–10 pairs ..... 25. **O. linearibracteata** P.C. Li.

21. Inflorescence densely capitate, not elongated; corolla 9–12 mm long. Pods oblong, (10) 15–20 (25) mm long with straight beak ...
.... 19. **O. humifusa** Kar. et Kir.

+ Inflorescence loosely capitate, highly elongated; corolla 5–6 mm long. Pods orbicular, 3–5 mm long, with uncinate beak .... 14. **O. chantengriensis** Vass.

22. Pods pendent or strongly declinate, on up to 5–7 mm long stalk ....23.

+ Pods erect, sessile or on short stalk, with very long beak ....27.

23. Inflorescence greatly elongated up to 5–12 cm at anthesis, lax. Pods orbicular or oval, asymmetrical, with short uncinate beak. Plant somewhat ash-grey ....24.

+ Inflorescence dense, capitate or oval, not significantly elongated. Pods oblong with long recurved beak. Plant canescent ....26.

24. Plant small (peduncles up to 10 cm tall), usually forming dense cushions; leaflets 5–6 pairs. Corolla 5–6 mm long; cusp of keel less than 1 mm long ....14. **O. chantengriensis** Vass.

+ Plant very large (peduncles up to 25 cm tall), not forming cushions; leaflets 6–12 pairs. Corolla 8–12 mm long; cusp of keel 1.5–2.5 mm long ....25.

25. Surface shoots covered with closely overlapping dead broad stipules. Corolla light-violet, usually turning yellow on drying .. .. 20. **O. imbricata** Kom.

+ Surface shoots without sheath of leaf remnants. Corolla bright-violet or sometimes white .... 28. **O. merkensis** Bge.

26. Leaflets 6–10 pairs. Corolla 7–8 mm long; cusp of keel about 3 mm long, curved. Pods oblong, 8–15 mm long, pendent .... 30 **O. globiflora** Bge.

+ Leaflets 12–18 pairs. Corolla 10–12 mm long; cusp of keel up to 1 mm long, straight. Pods oval, 7 mm long, horizontally declinate .... 17. **O. dumbedanica** Grub. et Vass.

27. Inflorescence racemose, oblong, up to 5 cm long, lax, highly elongated. Cusp of keel 2–3 mm long ....28.

+ Inflorescence capitate, oval, usually dense, not or poorly elongated. Cusp of keel up to 2 mm long ....29.

28. Corolla blue, 12–15 mm long, calyx 5 mm long. Leaflets glabrous, 7–15 mm long. Pod 15–25 mm long and 5–6 mm broad .... 15. **O. coerulea** (Pall.) DC.

9 + Corolla blue-violet or purple, 6–7 mm long, calyx 2.5–3 mm long. Leaflets with appressed hairs, 4–6 mm long. Pod 8–10 mm long and 4–5 mm broad .... 18. **O. filiformis** DC.

29. Corolla violet-coloured, 10–15 mm long. Leaflets 12–18 pairs ....30.

+ Corolla pale-yellow, 12–15 mm long, with violet-coloured keel. Leaflets 5–9 pairs .... 23. **O. ladyginii** Kryl.

30. Peduncles generally not exceeding leaves, 15–18 cm long, thickened. Corolla pale-violet, 10–12 mm long ....
.... 22. **O. kumbelica** Grub. et Vass.

+ Peduncles twice longer than leaves, up to 30 cm tall, slender. Corolla dark-violet, 13–15 mm long. Pods membranous, oval, transversely rugose ....24. **O. larionovii** Grub. et Vass.

31. Plant with distinct stem (section 3. **Mesogaea** Bge.) ....32.

+ Plant acauline or subacauline (section 1. **Oxytropis**)................50.

32. Flowers yellow. Stems well-developed, erect, villous ....46.

+ Flowers violet, blue or sky-blue. Stems developed or scarcely visible; if developed and erect, with appressed hairs.................33.

33. Stems well-developed, flexuose and branched, erect or procumbent; stipules interconnate. Flowers 5–10 mm long, small ....34.

+ Stems generally short or poorly visible; more often, only their uppermost internodes developed while rest are considerably shortened, ascending, procumbent or erect; stipules free or interconnate up to 1/2. Flowers 10–15 mm long, very large; if smaller, in dense capitate inflorescences, nutant. Inflorescence capitate or umbellate ....38.

34. Plants small with slender procumbent stems. Racemes dense, shortened, subcapitate. Pods 5–15 mm long ....35.

+ Plant usually large, with erect or ascending strong stem up to 60 cm tall. Racemes lax, elongated, often with distant flowers. Pods 10–30 mm long .... 39. **O. glabra** (Lam.) DC.

35. Pods with fine appressed hairs. Leaflets 6–15 pairs, with appressed hairs ....36.

+ Pods pubescent with patent long white hairs. Leaflets 4–8 pairs, with patent white hairs on both sides. Peduncles as long as leaves or longer ....36. **O. chorgossica** Vass.

36. Plants green, faintly pubescent. Peduncles as long as or longer than leaves ....37.

+ Plants canescent, with short and stiff white hairs. Peduncles shorter than leaves. Leaflets 7–9 pairs ....
.... 42. **O. hirsutiuscula** Freyn.

37. Leaflets 5–9 pairs. Peduncles 1.5–2 times longer than leaves. Standard about 8 mm long, cusp of keel 1.5–2 mm long; teeth of calyx as long as its tube ....
.... 39. **O. glabra** var. **salina** (Vass.) Grub.

+ Leaflets 10–15 pairs. Peduncles as long as leaves. Standard 8–10 mm long; cusp of keel up to 0.5 mm long; teeth or calyx 1/3–1/2 as long as tube .... 40. **O. gorbunovii** Boriss.

38. Plant acauline but with long subsurface shoots forming small, 3–5 cm thick, loose mats. Inflorescence capitate, 4–10-flowered, on

long peduncle. Standard narrowed at centre, panduriform, 14–17 mm long, keel with long (3–4 mm) cusp. Pods cystiform, 25–30 mm long and 15–20 mm broad. Leaflets 8–14 pairs .... 50. **O. platonychia** Bge.

10   +   Plants without long subsurface shoots but with generally distinct stems. Standard not narrowed ....39.

39.   Plants densely pubescent, ash-grey or white-villous, with procumbent stems and 2–6 mm long small leaflets ....40.

+   Plants green or greyish-green, sparsely pubescent, with erect or ascending stems ....41.

40.   Plants with semiappressed hairs, ash-grey. Leaflets 4–6 pairs; stipules interconnate .... 35. **O. cana** Bge.

+   Plants white-villous. Leaflets 8–11 pairs; stipules almost not connate .... 34. **O. cachemiriana** Camb.

41.   Stems and rachis with long divaricate hairs or villous ....42.

+   Stems and rachis with appressed hairs ....44.

42.   Inflorescence many-flowered, dense raceme, elongating at anthesis and in fruit; flowers nutant. Leaflets 15-25 pairs, somewhat declinate toward leaf base; stipules free. Plant up to 20 cm tall ....37. **O. deflexa** (Pall.) DC.

+   Inflorescence 5–8-flowered, capitate or umbellate, not elongating; flowers not nutant. Leaflets 6–14 pairs, perpendicular to rachis; stipules connate for 1/2. Plant up to 10 cm tall ....43.

43.   Leaflets 10–14 pairs, oval, 5–10 mm long; calyx long. Corolla 10–14 mm long. Pods vesicular, oval, about 30 mm long, with short patent and black hairs. Stems short, poorly developed ....51. **O. ulzijchutagii** Sancz.

+   Leaflets 6–9 pairs, oblong, 4–7 mm long and about 3 mm broad. Corolla 9–10 mm long. Pods oval, 10–12 mm long, glabrous. Stems generally developed, ascending ....38. **O. gerzeensis** P.C. Li.

44.   Stems slender, weak, lower internodes very short and only uppermost often developed; stipules ovate-lanceolate. Pods oval with short beak, pendent ....45.

+   Stems generally well-developed, strong, up to 30 cm tall; stipules lanceolate, herbaceous. Pods cylindrically oval, 15–25 mm long, with long uncinate beak, with patent white hairs, erect at first, declinate later. Plant very densely pubescent, grey-green ....41. **O. gueldenstaedtioides** Ulbr.

45.   Pods cylindrical, 8–15 mm long, with broad ventral furrow and short and appressed black hairs .... 44. **O. lapponica** (Wahl.) Gay.

+ Pods compressed flat on valve side, 15–20 mm long and 5–10 mm broad, keeled, without furrows, pubescent with short appressed white hairs; mature fruits calvescent ...... **46. O. melanocalyx** Bge.

46. Base of calyx, inflated specially in fruit, with long silky hairs. Inflorescence oblong-cylindrical. Pod oval, 12–15 mm long and 6–7 mm broad. Stem somewhat thick, strong ....
.... 47. **O. ochrocephala** Bge.

+ Calyx not inflated. Inflorescence capitate or ovate. Pod oblong or fusiform, 15–30 mm long. Stem slender, weak .... 47.

47. Plant green; stipules connate up to 1/2–2/3. Standard 15–18 mm long. Pod up to 20–30 mm long and 5–6 mm broad .... 48.

+ Plants villous with white hairs; stipules free. Standard 12–14 mm long. Pod 15–20 mm long and about 3 mm broad ....
.... 49. **O. pilosa** (L.) DC.

48. Tip of keel violet-coloured. Lower internodes of stem generally very short and only upper ones developed ....
.... 48. **O. ochroleuca** Bge.

+ All parts of corolla yellow. Stem normally developed .... 49.

49. Stipules herbaceous-membranous. Calyx with brownish pubescence; standard of corolla deeply emarginate at tip. Pod up to 20–30 mm long, patently pilose .... 45. **O. meinshausenii** C.A. Mey.

+ Stipules herbaceous, green. Calyx with short black hairs; standard slightly emarginate at tip. Pod 14–16 mm long, with appressed hairs... 43. **O. kansuensis** Bge.

50. Cushion-shaped plant with numerous lignifying shoots and very small oval leaflets, 1.5–3 mm long and 0.5–1.5 mm broad, 7–12 pairs. Inflorescence capitate, 3–6-flowered; corolla 8–10 mm long .... 10. **O. savellanica** Bge. ex Boiss.

+ Plants not forming cushions; leaflets very large .... 51.

51. Plants green, weakly pubescent or subglabrous, with long procumbent subsoil shoots. Flowers 2–5 in umbellate inflorescence, sometimes single .... 52.

+ Plant densely pubescent, greyish or canescent, leaflets with dense silky hairs; forming lax mats. Flowers numerous, in dense capitate inflorescences; if inflorescence 4–5-flowered, stipules of developed leaves glabrous .... 54.

52. Peduncles 1–3-flowered, with appressed hairs; calyx with appressed black hairs. Pod stalked, with appressed white hairs. Stipules white with hairs... 53.

+ Peduncle 3–5-flowered, with patent hairs; calyx with patent white hairs. Pod sessile, with appressed black hairs. Stipules glabrous, sparsely ciliate only along margin .... 8. **O. platysema** Schrenk.

53. Corolla white, 7–9 mm long; inflorescence usually 2-, rarely 3-flowered .... 1. **O. biflora** P.C. Li.

11

+ Corolla purple-coloured, 12–15 mm long; inflorescence usually 3-, rarely 1- or 2-flowered .... 7. **O. pauciflora** Bge.

54. Inflorescence many-flowered, capitate. Stipules with dense appressed hairs ....55.

+ Inflorescence umbellate, 4–5-flowered. Stipules diffusely hairy only at first, glabrous thereafter; ciliate only along margin ....4. **O. latialata** P.C. Li.

55. Petiole short, much shorter than blade. Pod with less than 1 mm long short beak, oblong or subglobose ....56.

+ Petiole very long, twice longer than blade or as long. Pod with uncinate beak, 1–2 mm long, ovate or oblong-ovate ....5. **O. lehmannii** Bge.

56. Standard and keel covered with dense silky pubescence on outer surface ....57.

+ All parts of corolla glabrous ....58.

57. Leaflets oblong or elongated-lanceolate, 8–25 mm long, 6–15 pairs. Peduncles longer than leaves; inflorescence somewhat lax, racemose; calyx shorter than corolla ....11. **O. sericopetala** C.E.C. Fisch.

+ Leaflets ovate-lanceolate or lanceolate, 8–12 mm long, 3–7 pairs. Peduncles shorter than leaves; inflorescence capitate, dense; calyx longer than corolla .... 6. **O. parasericopetala** P.C. Li.

58. Plant acauline or subacauline; leaflets 6–9 pairs; stipules membranous but their free portion herbaceous, ovate-lanceolate. Pods subglobose, lanate ....59.

+ Plant with short but distinctly developed stem, grey-silky; leaflets 3–4 (5) pairs; stipules membranous, their free portion shortly deltoid with thick midvein. Pods oblong, with appressed hairs. Keel with short recurved cusp .... 12. **O. tatarica** Camb.

59. Keel with long subulate, often uncinate, cusp. Plant grey-silky, with erect peduncles ....9. **O. proboscidea** Bge.

+ Keel with short, subconical cusp. Plant greyish-woolly, with procumbent or ascending peduncles .... 2. **O. glacialis** Benth.

60(7) Leaflets paired ....61.

+ Leaflets in whorls of 3 or more each ....121.

61. Plants densely covered with numerous adhesive sessile glands. Calyx of dead leaves rigescent and albescent ....62.

+ Plants without adhesive glands ....63.

62. Plants without stems or penduncles; flowers sessible at base of leaves. Pods hard, nutlet-shaped (section 10. **Leucopodia** Bge.) .. .. 143. **O. squamulosa** DC.

+ Plants acauline but with developed peduncles. Pods thin-walled, vesicular. Forming dense, cushion-shaped mats (section 9. **Gloeocephala** Bge.) .... 142. **O. fragilifolia** Ulzij.

12

63. Plants with developed stems or acauline. Pods elongated, oblong-oval or cylindrical-oblong ....64.
+ Plants acauline. Pods ovate or asymmetrical-oval ....99.
64. Plants acauline. Pods membranous, oblong-oval, or coriaceous, cylindrical-oblong ....76.
+ Plants small, with slender stems, 15–20 cm long, or acauline, with small, up to 10 mm long, leaflets; if longer, 1–2 mm broad, narrow. Pods coriaceous, oval or oblong. Calyx campanulate (section 4. **Ortholoma** Bge.) ....65.
65. Pods puberulent ....66.
+ Pods with long villous with white hairs ....73.
66. Stems slender, dichotonously-branched, weakly pubescent, 10–20 cm tall. Pods stalked, pendent, oblong, 15–20 mm long. Inflorescence oval ....86. **O. podoloba** Kar. et Kir.
+ Characteristics different ....67.
67. Racemes oblong, lax, with distant lower flowers. Pods erect ... 68.
+ Inflorescence short, capitate or ovate ....69.
68. Plant ash-grey due to dense white pubescence; stems ascending or procumbent, slender, numerous; leaflets 8–12 pairs. Standard emarginate at tip. Pods oblong, 15–20 (25) mm long ....
.... 82. **O. floribunda** (Pall.) DC.
+ Plant green, weakly pubescent; stems erect; leaflets 5–8 pairs. Standard rounded at tip. Pods linear-oblong, about 10 mm long
....90. **O. tenuis** Palib.
69. Pods pendent, short, 7–10 mm long. Plant with appressed white hairs; stems slender, branched; leaflets 5–8 pairs. Standard 8–9 mm long .... 88. **O. sarkandensis** Vass.
+ Pods erect or declinate but not pendent ....70.
70. Stems lignified at base, procumbent. Corolla 9–12 mm long, cusp of keel 0.5–1 mm long ....71.
+ Plants acauline or with very short (1–3 cm long) herbaceous stem, ash-grey due to dense pubescence. Corolla 11–15 mm long; cusp of keel 2–3 mm long ....72.
71. Inflorescence 2–3-flowered; corolla 12 mm long; wings bilobed at tip .... 83. **O. fruticulosa** Bge.
+ Inflorescence capitate; 5–10-flowered; corolla 9–12 mm long; wings undivided at tip. Pods oblong-oval, 10–15 mm long, with short with patent white hairs .... 91. **O. tianschanica** Bge.
72. Standard of corolla deeply emarginate at tip, bilobed; caylx tubular-campanulate, 7–8 mm long, with dense villous with white hairs ....79. **O. biloba** Sap.
+ Standard of corolla slightly emarginate at tip; calyx campanulate, 5–7 mm long, densely covered with brown hairs ....
.... 89. **O. schrenkii** Trautv.

13

73. Stems straight, branched. Inflorescence racemose, with distant flowers ....74.

+ Plants forming dense mats or cushions. Inflorescence capitate or umbellate ....75.

74. Leaflets 5–9 pairs. Plant puberulent with white hairs. Corolla 12–13 mm long ....85. **O. hirsuta** Bge.

+ Leaflets 3–4 pairs. Plant canescent, appressed-pilose. Corolla 15–17 mm long .... 84. **O. grum-grshimailoi** Palib.

75. Plant densely-caespitose, tomentose with white hairs; leaflets 8–14 pairs. Inflorescence capitate, many-flowered; corolla pink-violet; standard 9–12 mm long .... 81. **O. dichroantha** Schrenk.

+ Plant cushion-shaped, sericeous with ash-grey hairs; leaflets 5–8 pairs. Inflorescence umbellate, 3–4-flowered; corolla light-blue, standard 8 mm long .... 87. **O. pulvinoides** Vass.

76. Pod membranous, oblong or oblong-oval, erect. Calyx tubular or tubular-campanulate; ovary sessile (section 3. **Orobia** (Bge.) Aschers. et Graebn.) ....77.

+ Pod coriaceous, cylindrical-oblong up to linear-oblong, erect. Calyx shortly-campanulate, rarely tubular-campanulate but then ovary stalked (section 1. **Eumorpha** Bge.) ....94.

77. Corolla purple, violet, blue, pale-violet or pale-pink ....78.

+ Corolla pale-yellow, 17–20 mm long. Racemes short, dense. Plant small (8–15 cm tall), with dense silvery-hairs ....
.... 75. **O. recognita** Bge.

78. Cusp of keel 3–4 mm long. Pods 8–10 mm broad. Leaflets 5–8 or 8–15 pairs ....79.

+ Cusp of keel short (0.5–1 mm); if 1–2 mm long, leaflets 10–24 pairs or racemes long, spicate. Pods 4–7 mm broad ....80.

79. Corolla 25–35 mm long. Pods appressed-pilose. Leaflets 5–8 pairs
.... 69. **O. grandiflora** (Pall.) DC.

+ Corolla 20–25 mm long. Pods with soft patent hairs. Leaflets 8–15 pairs .... 67. **O. frigida** Kar. et Kir.

80. Inflorescence elongated, racemose, with distant flowers in lower part. Leaflets 18–24 pairs ....81.

+ Inflorescence capitate or umbellate. Leaflets 5–16 (18) pairs ...82.

81. Leaflets green, subglabrous. Bracts 3–5 mm long; calyx tubular, 8–10 mm long .... 66. **O. confusa** Bge.

+ Leaflets densely appressed-pilose on both surfaces, canescent-sericeous. Bracts 7–12 mm long; calyx tubular-campanulate, 9–12 mm long .... 76. **O. soongorica** (Pall.) DC.

82. Standard of corolla deeply emarginate at tip, bilobed ....83.

+ Standard rounded at tip or barely emarginate ....86.

83. Racemes many-flowered, dense. Plant green ....84.

14

+ Racemes 3–6-flowered, lax. Plant densely sericeous with white hairs .... 65. **O. chionophylla** Schrenk.
84. Plant large, 15–30 cm tall. Leaflets 10–25 mm long ....85.
+ Plant 10–15 cm tall. Leaflets 5–12 mm long, younger ones densely sericeous with white hairs, later glabrant; green on upper surface .... 61. **O. alpina** Bge.
85. Leaflets glabrous or diffusely pilose, green. Pods with patent hairs. Plant 15–20 cm tall ....62. **O. altaica** (Pall.) DC.
+ Leaflets with appressed white hairs on both surfaces; later glabrescent and green. Pods appressed-pilose. Plant 15–30 cm tall .... 63. **O. ambigua** (Pall.) DC.
86. Inflorescence many-flowered, dense, capitate or oval, elongating after anthesis; corolla 17–25 mm long. Leaflets 7–15 (20) mm long. Plant quite large, up to 25 cm tall ....87.
+ Inflorescence few-flowered, somewhat lax; corolla generally 11–17 mm long, more rarely 22–25 mm long. Leaflets 6–8 mm long and 2–4 mm broad. Plants smaller, low, 10–15 cm tall ....90.
87. Plants green; only younger leaflets with dense appressed hairs, older ones subglabrous, green on upper surface. Inflorescence not pendent ....88.
+ Plant with dense silvery hairs. Young inflorescence often pendent ....64. **O. argentata** (Pall.) Pers.
88. Inflorescence tufted due to long, nearly equal bracts. Plant diffusely hairy, green ....72. **O. longibracteata** Kar. et Kir.
+ Inflorescence not tufted; bracts not longer than calyx. Plants initially with appressed silky hairs, later glabrescent and greenish ....89.
89. Leaflets 6–7 (11) pairs. Peduncles puberulent; corolla 20–25 mm long .... 71. **O. latibracteata** Jurtz.
+ Leaflets 9–12 (15) pairs. Peduncles with long patent hairs; corolla 18–20 mm long ....77. **O. strobilacea** Bge.
90. Calyx teeth 1/6–1/3 of tube...91.
+ Calyx teeth as long as tube or only 1/2 as long; linear-lanceolate. Inflorescence umbellate, lax, 3–6-flowered .... 78. **O. tschujae** Bge.
91. Corolla 22–25 mm long; calyx tubular, 15–17 mm long. Inflorescence 3–8-flowered ....73. **O. macrosema** Bge.
+ Corolla 11–17 mm long; calyx tubular-campanulate, 6–10 mm long ....92.
92. Inflorescence capitate, later elongated. Pods appressed-pilose .....
....93.
+ Inflorescence umbellate, 4–8-flowered. Pods with patent hairs ....
....70. **O. ketmenica** Sap.

93. Calyx with patent hairs; corolla 11–13 mm long; cusp of keel 0.5–0.75 mm long .... 68. **O. gebleri** Fisch. ex Bge.

+ Calyx with appressed hairs; corolla 12–17 mm long; cusp of keel 1–1.5 mm long .... 74. **O. martjanovii** Kryl.

94. Leaflets 18–25 pairs. Corolla white or yellow, sometimes with bright-violet keel, 16–20 mm long; racemes many-flowered, lax, highly elongated. Pods 20–30 (40) mm long; compressed-cylindrical .... 55. **O. macrocarpa** Kar. et Kir.

+ Leaflets 10–16 pairs. Corolla violet or purple; inflorescence capitate or oval, more rarely laxly racemose, not elongated. Pods 12–25 mm long ....95.

95. Inflorescence laxly racemose, with distant flowers; peduncles shorter than leaves. Corolla 9–13 mm long. Leaflets linear-lanceolate, 11–13 pairs. Pods oblong, 18–25 mm long ....
.... 56. **O. semenovii** Bge.

+ Inflorescence capitate or oval; peduncles longer than leaves; corolla 15–18 mm long. Leaflets ovate to broadly-lanceolate, 10–16 pairs ....96.

96. Leaflets 12–16 pairs. Standard 15–17 mm long; calyx 6–10 mm long. Pods oblong, 12–15 mm long ....97.

+ Leaflets 10–12 pairs. Standard 17–18 mm long; calyx 10–12 mm long. Pods linear-cylindrical, 20–25 mm long ....
.... 53. **O. cuspidata** Bge.

97. Plants green; leaflets with sparse or diffuse appressed hairs. Peduncles 2–3 times longer than leaves, 20–30 cm tall; standard 15 mm long; calyx 5–8 mm long ....98.

+ Plant greyish-green, densely pubescent; leaflets with densely sericeous hairs. Peduncles 12–18 cm tall, up to twice longer than leaves; standard 17 mm long; calyx 10 mm long, teeth only slightly shorter than tube .... 52. **O. barkultagi** Grub. et Vass.

98. Peduncles appressed-pilose; standard with orbicular limb; calyx 7–8 mm long, teeth slightly shorter than tube ....
.... 54. **O. dschagastaica** Grub. et Vass.

+ Peduncles with patent hairs; standard lingulately elongated, with narrow incision at tip; calyx 5–6 mm long, teeth 1/2–2/5 as long as tube .... 57. **O. taldycola** Grub. et Vass.

99 (63). Flowers very small; 6–13 mm long, dense heads on peduncles. Pods ovate, stellately spreading villous with long hairs. Petioles slender, withering (section 2. **Sphaeranthella** Gontsch.) ....100.

+ Flowers large, 20–30 mm long. Pods globose-ovate, not forming stellately spreading head. Petioles of dead leaves persistent and rigescent (section 5. **Xerobia** Bge.) ....102.

100. Leaflets 5–8 (10) pairs. Calyx teeth shorter than tube or as long . ...101.

+ Leaflets 8–15 pairs, lanceolate, velutinous. Calyx teeth 1.5 times longer than tube; standard 7.5 mm long .... 60. **O. valerii** Vass.

101. Calyx teeth 2/5 as long as tube; calyx 4–5 mm long; standard 6–7 mm long. Leaflets oval, white-tomentose; rachis and peduncle thickened. Pods 8–10 mm long, with 3–4 mm long, distinctly marked beak .... 59. **O. crassiuscula** Boriss.

+ Calyx teeth as long as tube; calyx 7–8 mm long; standard 11–13 mm long. Rachis and peduncle slender; leaflets linear or oblong, with dense appressed hairs, ash-grey. Pods 12–15 mm long, with short beak .... 58. **O. caespitosula** Gontsch.

16 102. Pods hard, nutlet-shaped, with thick coriaceous walls, tomentose-pubescent ....103.

+ Pods vesicular, with thin membranous walls, pubescent or glabrous ....111.

103. Leaflets 3; rarely 5 but then leaflets linear and tufted or leaves simple, with single leaflet. Inflorescence 2–4-flowered, or flower single ....104.

+ Leaflets 5 or more. Inflorescence few-flowered, lax, or many-flowered, dense ....106.

104. Leaves simple, with single lanceolate leaflet. Inflorescence 1–4-flowered .... 104. **O. monophylla** Grub.

+ Leaves with 3–5 leaflets ....105.

105. Leaflets linear-lanceolate, 3, but early leaves with very small elliptical leaflets. Inflorescence 1–2-flowered .... 96. **O. diversifolia** Peter-Stib.

+ Leaflets filamentous-linear, 3, more rarely 5, forming tuft. Peduncles 1-flowered .... 99. **O. junatovii** Sanczir.

106. Inflorescence many-flowered, capitate, dense; bracts long, as long as calyx or its tube. Leaflets lanceolate or linear-lanceolate, glabrous, fimbriate along margin .... 110. **O. setosa** (Pall.) DC.

+ Inflorescence 2–6-flowered, umbellate, lax; bracts shorter than calyx tube ....107.

107. Leaflets densely pilose on both surfaces, grey ....108.

+ Leaflets glabrous on upper or on both surfaces, green ....110.

108. Leaflets with patent hairs ....109.

+ Leaflets with dense appressed hairs, sericeous, frequently longitudinally folded. Peduncles shorter than leaves but distinct .... 97. **O. eriocarpa** Bge.

109. Inflorescence subsessile; calyx teeth 1/5–1/4 of tube; corolla violet-coloured. Leaflets elliptical, softly hairy .... 108. **O. rhizantha** Palib.

+ Inflorescence on distinct peduncle, shorter than leaves or as long; calyx teeth only 1/2 of tube; corolla yellow. Leaflets elongated-lanceolate, strigose .... 109. **O. setifera** Kom.

110. Leaflets linear, 10–25 mm long and 1–2 mm broad, 2, rarely 3 pairs. Clayx 12–13 mm long, teeth 1/5–1/4 of tube; standard 22–24 mm long .... 100. **O. klementzii** Ulzij.

+ Leaflets oval, 5–10 mm long and 3–5 mm broad, 3–6 pairs. Calyx 14–17 mm long; teeth 1/2 of tube, linear-lanceolate; standard 25–30 mm long ....98. **O. intermedia** Bge.

111. Leaflets glabrous or very sparsely pubescent above and plants green ....112.

+ Leaflets with dense soft hairs on both surfaces and plant ash-grey or greyish due to pubescence ....115.

112. Flowers ivory-white, violet spot only on tip of keel. Leaflets fleshy, glabrous on both surfaces but ciliate along margin. Pod glabrous ....95. **O. ciliata** Turcz.

+ Flowers purple or violet. Leaflets appressed-pilose beneath, without cilia along margin. Pod pilose ....113.

113. Leaflets linear or filamentous-linear, up to 40 mm long ....114.

+ Leaflets dimorphic-oval on some leaves, 5–10 mm long and 2.5–4 mm broad; of other, much later ones, oblong, 10–20 mm long but of same breadth as above. Racemes lax, 5–8-flowered .... 103. **O. mixotriche** Bge.

17 114. Leaflets 3–4 pairs, sericeous on upper surface. Inflorescence many-flowered, capitate .... 102. **O. micrantha** Bge. ex Maxim.

+ Leaflets 4–6 pairs, glabrous on upper surface. Inflorescence lax, 2–5-flowered..................................... 101. **O. leptophylla** (Pall.) DC.

115. Racemes many-flowered (5–7 or more); bracts broad and long, as long as or 1/2 of calyx ....116.

+ Racemes with (1) 2–3 (6) flowers; bracts very short and narrow . ...118.

116. Leaflets large, oblong, 10–15 mm and up to 30 mm long, silvery-sericeous. Plant largely-caespitose, up to 15 cm tall .... 105. **O. nitens** Turcz.

+ Leaflets very small. Plants finely-caespitose, usually up to 10 cm tall ....117.

117. Peduncles longer than leaves, erect. Plants loosely-caespitose .... 107. **O. pseudofrigida** Sap.

+ Peduncles shorter than leaves, procumbent or ascending. Plants densely-caespitose .... 93. **O. assiensis** Vass.

118. Petioles lignified in lower part, brittle only at tip, leaflets 4–5 pairs. Racemes 3–6-flowered ....106. **O. potaninii** Bge. ex Palib.

+ Petioles not lignified, brittle; leaflets (3) 5–8 pairs. Racemes (1) 2–4-flowered ....119.

119. Leaflets densely appressed-pilose and entire plant silvery-sericeous. Pod lanate with short hairs ....120.

+ Leaflets densely appressed-pilose on both surfaces and entire plant villous .... 92. **O. ampullata** (Pall.) Pers.

120. Leaflets 7–9 (12) mm long, 7–8 pairs. Minute glandular hairs seen in pubescence of calyx and pods ....94. **O. burchan-buddae** Grub. et Vass.

+ Leaflets 3–7 (10) mm long, 3–5 (6) pairs. Glandular hairs absent in pubescence .... 111. **O. stracheyana** Benth. et Baker.

121(60) Plant with sessile glands or glandular hairs (section 8. **Polyadena** Bge.) ....147.

+ Plant without sessile glands or glandular hairs ....122.

122. Calyx tubular or tubular-campanulate (section 6. **Baicalia** Bge.) ... .123.

+ Calyx short-campanulate; flowers small, 7–11 mm long (section 7. **Gobicola** Bge.)..146.

123. Plants acauline or with shortened stem, inflorescence terminal.... ....124.

+ Plants with well-developed branched stems; leaves with 5–6 whorls of 4 linear leaflets each. Flowers single or 2–3 together, axillary ....130. **O. ramosissima** Kom.

124. Flowers violet, purple, pink or blue, rarely white, but wings sometimes yellow or greenish. All leaves with whorled leaflets .. ..125.

+ Flowers yellow. Early leaves pinnate (leaflets paired), later ones with whorled leaflets .... 122. **O. ochrantha** Turcz.

125. Stipules with dense white tomentum; as a result, base of petiole and tip of caudex densely wrapped in white tomentum ....126.

+ Stipules glabrous or pilose but not tomentose-pubescent ....129.

126. Leaflets with dense white or grey tomentum, in closely set whorls ....127.

+ Leaflets with sparse appressed hairs, in distant whorls ....128.

127. Caudex elongated; leaflets with grey tomentum, oblong or linear, 4–8 each in 12–18 whorls........................ 116. **O. lanata** (Pall.) DC.

+ Caudex very short; leaflets with white tomentum, lanceolate or ovate, obtuse, 3–7 each in 10–15 whorls.................................... ....118. **O. lanuginosa** Kom.

128. Leaflets oval or ovate, 3–6 each in 20–30 unevenly distant whorls .... 120. **O. mongolica** Kom.

+ Leaflets oblong or linear, fleshy, with convoluted edges, 4 each in 7–15 evenly distant whorls ....119. **O. lasiopoda** Bge.

129. Leaflets linear-subulate, generally 8 together in 25–30 whorls .... 121. **O. myriophylla** (Pall.) DC.

+ Leaflets 3–6 in each of 4–15 (18) whorls....................................130.

18

130. Leaves with 2–6 whorls of linear or lanceolate leaflets .... 131.

\+ Leaves with (5) 6–18 whorls of leaflets .... 133.

131. Corolla violet-coloured. Leaves with 2–5 whorls of very small leaflets. Plants 3–6 cm tall .... 132.

\+ Standard and wing of corolla yellow-green and keel violet-coloured at tip. Leaves with 5–6 whorls of leaflets .... 135. **O. viridiflava** Kom.

132. Inflorescence capitate, 7–12-flowered; corolla about 15 mm long. Pod globose-ovate, 12–13 mm long . 129. **O. pumila** Fisch ex DC.

\+ Inflorescence 1–3-flowered; corolla 21–23 mm long. Pod oblong-ovate, 16–23 mm long .... 117. **O. langshanica** H.C. Fu.

133. Plant covered with stiff straight patent hairs. Flowers 25–30 mm long; cusp of keel very long, 3–5 mm .... 114. **O. fetissovii** Bge.

\+ Plant softly sericeous or tomentose, more rarely subglabrous. Flowers smaller, up to 25 mm long; cusp of keel 0.5–2 mm long. .... 134.

134. Flowers (1) 2–5 (8) in umbellate inflorescence. Leaves densely sericeons or densely lanate with silvery hairs .... 135.

\+ Flowers many, in capitate or racemes inflorescence .... 141.

135. Leaflets very small, 1–3 mm, rarely up to 5 mm long, in 6–12 closely set whorls. Inflorescence 1–3-flowered; peduncles scarcely exceeding leaves. Small densely-caespitose plant .... 136.

\+ Leaflets very large, 3–8 mm long, in 5–8 distant whorls. Peduncles exceeding leaves, with 3–6 (8) flowers .... 138.

136. Plant with surface cauline shoots. Flowers up to 20–22 mm long .... 126. **O. pellita** Bge.

\+ Plant without surface shoots. Flowers 10–16 mm long .... 137.

137. Leaflets 1–1.5 mm long, in 6–10 whorls. Peduncles with scattered minute glands, with 2, more rarely 1 or 3, flowers; corolla 10–13 mm long .... 134. **O. sutaica** Ulzij.

\+ Leaflets 2–3 mm long, in 10–12 whorls. Peduncles without glands, with 1–3 flowers; corolla 13–16 mm long .... 113. **O. chionobia** Bge.

138. Flowers 21–25 mm long. Pods large, globose-inflated, 25 mm long and 20 mm broad, scarious .... 131. **O. rhynchophysa** Schrenk.

\+ Flowers 10–17 mm long. Pods very small, membranous .... 139.

139. Peduncles straight, erect. Calyx pubescent with white and black hairs; corolla 13–17 mm long, bright-violet or white .... 140.

\+ Peduncles procumbent-ascending, with 5–8 flowers. Clayx pubescent with only white hairs; corolla 10–12 mm long, light-violet or albescent .... 133. **O. selengensis** Bge.

140. Corolla bright-violet, 13–15 mm long. Peduncles sericeous like bracts and calyx. Flowers generally 3 .... 123. **O. oligantha** Bge.

+ Corolla white, yellowish or bluish on drying, 16–17 mm long. Peduncles with long patent hairs, bracts and calyx villous. Flowers mostly up to 8 .... 132. **O. saurica** Sap.

141. Leaflets linear-lanceolate to oblong-elliptical, appressed-pilose; if leaflets oval, pods coriaceous, hard; if pods membranous, small, up to 10 mm long .... 142.

+ Leaflets subcircular, oval or broadly-elliptical, 3–5 mm long, 3–6 each in 7–18 whorls but often paired, with fine white tomentum (like bracts and calyx). Inflorescence capitate, 5–10-flowered; corolla 22–25 mm long, violet-coloured. Pods membranous, subglobose, 16–17 mm long .... 128. **O. przewalskii** Kom.

142. Peduncles and leaves procumbent or ascending .... 143.

+ Peduncles and leaves straight, erect .... 145.

143. Leaflets fleshy, linear-oblong, glabrous on upper surface, sparsely pilose beneath. Flowers about 25 mm long. Pods coriaceous, glabrous, hard, oval. Peduncles shorter than leaves ....
.... 127. **O. prostrata** (Pall.) DC.

+ Leaflets not fleshy, appressed-pilose on both surfaces. Flowers 13–18 mm long. Pods pilose .... 144.

144. Inflorescence racemose, lax; peduncles longer than leaves. Pods coriaceous, hard, oblong-ovate, with deeply impressed furrow dorsally and ventrally .... 112. **O. bicolor** Bge.

+ Inflorescence capitate, dense; peduncles shorter than leaves or as long. Pods membranous, globose, without furrows ....
.... 125. **O. pavlovii** B. Fedtsch. et Basil.

145. Leaflets ash-grey, with densely sericeous, oblong or oval, 4–10 mm long, obtuse. Flowers about 25 mm long in lax inflorescence. Pods coriaceous, hard, oval, with white hairs ....
.... 115. **O. heterophylla** Bge.

+ Leaflets green, with sparse appressed hairs, linear-lanceolate, 10–20 (30) mm long, acute. Flowers very small, 13–18 (20) mm long, in capitate inflorescence. Pods scarious, vesicular-inflated, ovate, pilose or glabrous .... 124. **O. oxyphylla** (Pall.) DC.

146(122). Inflorescence globose, dense, many-flowered; calyx 6–7 mm long. Standard of corolla 9–11 mm long. Leaflets oblong, 4–6 each in 8–14 whorls .... 136. **O. gracillima** Bge.

+ Inflorescence lax, 5–9-flowered; calyx 5 mm long, standard 7–9 mm long. Leaflets linear-oblong, 4 each in 8–11 whorls ....
.... 137. **O. racemosa** Turcz.

147(121). Pod coriaceous, hard, oblong or fusiform straight or falcate ....
.... 148.

+ Pod scarious, vesicular-inflated, finely-tuberculate-glandular and scattered-pilose. Leaves with 12–15 whorls of leaflets. Peduncles with patent hairs, racemes dense .... 141. **O. trichophysa** Bge.

148. Caudex with white tomentum at tip. Peduncles and rachis villous with white hairs; leaflets oval or elliptical .... 149.

+ Plants caespitose, with numerous shortened, glabrous or appressed-pilose shoots. Peduncles and rachis appressed-pilose or glabrous; leaflets linear, developed, subglabrous .... 150.

149. Ovary glabrous; pod tuberculate-glandular, glabrous. Leaflets villous with white hairs on both surfaces, in 18–25 whorls of 4 each, rarely 6 each .... 140. **O. microphylla** (Pall.) DC.

+ Ovary appressed-pilose; pod diffusely tuberculate-glandular and diffusely pilose, sometimes subglabrous. Leaflets subglabrous on upper surface, with diffuse stiff-hairs and punctated glands beneath, in 14–20 whorls of 3–6 each 138. **O. chiliophylla** Royle.

150. Leaflets in whorls. Corolla lurid. Pod densely tuberculate-glandular, glabrous .... **O. muricata** (Pall.) DC.

+ Leaflets mostly paired, sometimes only in midportion of leaf and, in late leaves, whorled as well. Corolla purple-violet, sometimes white. Pod without tuberculate glands, appressed-pilose .... 139. **O. falcata** Bge.

151 (6) Calyx inflated in fruit; peduncles shorter than leaves, 1–2-flowered. Plant small and densely-caespitose; leaves with 1–2 pairs of leaflets (subgenus III. **Physoxytropis** Bge.) .... 140. **O. bungei** Kom.

+ Calyx not inflated; inflorescence many-flowered, on peduncles usually exceeding leaves. Leaflets multijugate .... 152.

152. Inflorescence capitate, crinite due to long teeth of calyx. Plant small, softly hairy, ash-grey due to pubescence (subgenus IV. **Ptiloxytropis** Bge.) .... 153.

+ Inflorescence spicate, long and dense. Large hirsute, lurid plant with tall peduncles (subgenus V. **Triticaria** Vass.) .... 148. **O. hirta** Bge.

153. Peduncles thick, strong, like rachis, rigescent, persistent, brittle; calyx teeth 1/2 of tube .... 146. **O. sacciformis** H.C. Fu.

+ Peduncles slender, like rachis, withering and impersistent; calyx teeth as long as tube or longer .... 154.

154. Calyx 6–7 mm long with teeth as long as tube; standard of corolla 7–8 mm long, cusp of keel up to 1 mm long .... 145. **O. bella** B. Fedtsch.

+ Calyx 7–9 mm long, teeth 1.5 times longer than tube; standard 9–10 mm long, cusp of keel about 2 mm long .... 147. **O. trichosphaera** Freyn.

Subgenus 1. **OXYTROPIS** (*Phacoxytropis* Bge.)

Section 1. **Oxytropis** (*Protoxytropis* Bge.)

1. **O. biflora** P.C. Li in Acta phytotax. sin. 18, 3 (1980) 369; id. in Fl. Xizang. (1985) 850. —Ic.: Fl. Xizang. 2, tab. 280, fig. 8–14.
Described from Tibet. Type in Beijing (PE).
In alpine meadows.

IIIB. Tibet: *Chang Tang* (Shuankhu—"Shuanghu, alt. 5000 m, on meadow, July 24, 1976—Qinghai-Xizang Compl. Exped. No. 1193, typus!").
General distribution: endemic.

2. **O. glacialis** Benth. ex Bge. in Mém. Ac. Sci. St.-Petersb., 7 sér. 22, I (1874) 18; ? Alcock. Rep. Pamir Boundary Commiss. (1898) 21; Hand.-Mazz. in Oesterr. bot. Z. 79 (1930) 33; Fl. Xizang. 2 (1985) 855 ex ptc., excl. syn. *O. proboscidea* Bge.—*O. nivalis* Franch. in Bull. Mus. nat. hist. natur. 3 (1897) 323; Grub. and Ulzij. in Novosti sist. vyssh. rast. 27 (1990) 93. —Ic.: Fl. Xizang. 2, tab. 281, fig. 15–21.
Described from South. Tibet. Type in St.-Petersburg (LE). Map 1.
On sandy-pebbly banks of brooks and rivulets, rocky slopes in alpine belt, 4000–5500 m alt.

IIIB. Tibet: *Chang Tang* (Mang-tze, dans la chaine de Oustoun-tagh alt. 5270 m, Sept. 6–7, 1892, D. de Rhins—typus *O. nivalis*!; "Quellfluss des Keria-Darja, enge Talsohle, 5200 m, July 6; Uferhügel, steinige Hange, westliche Teil des Sees Arpo-Zo, 5440 m, Aug. 24; Sumdschiling-Ebene, 5410 m, Aug. 5—1906, Zugmayer"—Hand.-Mazz. l.c.), *South* (Plains of Tibet, 15, 500 ft—[Aug. 1848] No. 3, Strachey and Winterb., typus!).
IIIC. Pamir ("in sandy soil along R. Aksu, 4000–4250 m, No. 17697" (Alcock l.c.).
General distribution: Himalayas (Kashmir).

3. **O. globiflora** Bge. in Mém. Ac. Sci. St.-Petersb. 7 sér., 14, 4 (1869) 43; ibid, 22, 1 (1874) 16; Vassilcz. and B. Fedtsch. in Fl. SSSR, 13 (1948) 17; Fl. Kirgiz. 7 (1957) 366; Fl. Kazakhst. 5 (1961) 340; Filim. in Opred. rast. Sr. Azii [Key to Plants of Mid. Asia] 7 (1983) 342; C.Y. Yang in Claves pl. Xinjiang, 3 (1985) 88. —*O. pagobia* Bge. in Mém. Ac. Sci. St. Petersb. 7 sér. 22, I (1874) 27; Danguy in Bull. Mus. nat. hist. natur. 19 (1913) 500; Hedin in S. Tibet, 6, 3 (1922) 62; Persson in Bot. notiser (1938) 291; Vassilcz. and B. Fedtsch. in Fl. SSSR, 13 (1948) 28; Fl. Kirgiz. 7 (1957) 370; Fl. Kazakhst. 5 (1961) 348; Ikonnik. Opred. rast. Pamira [Key to Plants of Pamir] (1983) 174; C.Y. Yang l.c. 83. —*O. lacostei* Danguy in Bull. Mus. nat. hist. natur. 14 (1908) 130. —*O. longialata* P.C. Li in Acta phytotax. Sin. 18, 3 (1980) 371; Fl. Xizang. 2 (1985) 862. —Ic.: Fl. Kazakhst. 5, Plate 48, fig. 1 and 3.
Described from Cen. Tien Shan. Type in St.-Petersburg. Plate I, fig. 3.
On rock screes, mossy-rocky tundra, *Cobresia* wastelands, moraines, rocky meadow and meadow-steppe slopes, coastal meadows and pebble beds of rivers in middle and upper mountain belts, up to 5000 m.

IB. Kashgar: *Nor.* (Kara-Teke mountain range, nor. slope, 1850–2450 m, June 7, 1889—Rob.; Uchturfan, Airi gorge, June 6; same site, Karagailik gorge, June 18, 1908 —Divn.), *West.* (Sarykol mountain range, Bostan-Terek, July 11, 1929—Pop "Jerzil, 2800 m, July 5, 1930; Bostan-Terek, about 2400 m, Aug. 4, 1934"—Persson l.c.; 8 km east of Kensu mine on road to Irkeshtam from Kashgar, 2200 m, June 17; 25 km nor.-west of Baikurt settlement along road to Torugart from Kashgar, along valley trail, June 20; 10 km south-east of Baikurt settlement along old road to Kashgar from Torugart, hummocky desert, June 21, 1959, Yun. and I-f. Yuan'), *South.* (Keriya mountain range, along Kizylsu river, 3050 m, July 24, 1885—Przew.; Karatash area, 3500–4000 m, Aug.-Sep. 1941—Serp.).

IIA. Junggar: *Tarb.* (Saur mountain range, upper B. Ob river, rocky tundra, June 13, 1904—Sap.), *Tien Shan* (Sairam, Talki brook, July 22; on Tekes river bank in midcourse, 1500 m, Aug. 13, 1877, A. Reg.; Sharysu river, 2450 m, June 21; nor. slope of Tekes valley, 1500 m, June 23; Agyaz river, 2150–2450 m, June 26; Bogdo mountain, 2450–2750 m, July 24; Sairam, Kyzemchek, 3050 m, July 31; Chubaty pass (Sairam), 2450–2750 m, Aug. 24, 1878, A. Reg.; Sairam-Nor valley, July 12; Urten-Muzart gorge, Aug. 3; Sharaboguchi, Aug.; Yuldus, Sep. 1878, Fet.; Taldy, 2450–2750 m, May 27; upper Taldy, 2750–3050 m, May 28; Kum-Daban, 2750 m, May 29; Kumbel', 2750–3050 m, May 30; same site, June 3; Bagaduslung-Bainamun, 1500–1850 m, June 4; Bagaduslung, eastern tributary of Dzhin, 2150–2750 m, June 4; Borgaty pass, 2450–2750 m, June 7; Nilki brook of Kash river, 2150 m, June 8; Borborogusun, 2750 m, June 15; south. tributary of Karagol, 2750 m, June 16; west. tributary of Aryslyn, 2450–2750 m, July 8; Aryslyn gorge, 2450–2750 m, July 10; Aryslyn, 2750 m, July 11, 1879, A. Reg.; "Sairam-Nor, montagnes, No. 955, July 23, 1895, Chaff."—Danguy, 1913, l.c.; Ishigart pass from Iirtash to Akshiirak, June 30, 1902—Sap.; Sabawtscho Tal bei der Alpe, June 22–24; oberes Kukurtuk Tal im Ansteig zum, June 26—July 2, 1903; im Dschanart und Kum-Aryk Tal—Merk Passe zwischen Kinsu und Kurdai, July 3; Musstamas Tal im sudlichen seiten Tal Aksai, Mitte, July 1907; sudliches Kiukonik Tal unter Tschon-Yailak Pass, June 15, 1908—Merzb.; 30 km west of Chzhaosu, Tszin'tsyuan', intermontane basin, No. 3267, Aug. 12, 1957—Kuan; Malyi Yuldus, [road] to Ulyastai, 2500 m, on alluvium, No. 6346, Aug. 2; Bain-Bulak, Khotun-Sumbul district [Bagrashkul' region], 2560 m, in valley on wet alluvium, No. 6394, Aug. 6, 1958, Lee and Chu; upper course of Khanga river 7–8 km before Kotyl' pass along road to Yuldus from Karashar, 2900 m, *Cobresia* wasteland, Aug. 1; Kotyl' pass, 3100 m, *Cobresia* patch, Aug. 15, 1958, Yun. and I-f. Yuan'; in Kucha town region, 2620 m, No. 10049, July 27, 1955—Lee and Chu.

IIIB. Tibet: *Weitzan* (mountains along Bochyu river, 4250 m, June 15; left bank of Yangtze river, 3950 m, June 23, 1884, Przew.).

IIIC. Pamir (" On moraines of Korumde glacier, Mustagh-Ata, 4367 m, July 27; Kara-Jilga valley and spring at Basik-Kul, 3727 m, July 24, 1894"—Hedin l.c.; "Bords de la riviere Beik, 4060 m, July 23, 1906, Lacoste"—Danguy, 1908 l.c.; around Ucha pass, June 16; Ulug-tuz gorge in Charlym river basin, June 23; same site, in meadow of brook, June 28, 1909, Divn.; crossing from Arpalyk river gorge to Kyzyl-Bazar, 3500 m, June 13, 1941; Kashkasu river source, 4200–5500 m, July 5; moraine waterdivide between Atrakyr and Tyuzutek rivers, 4500–5000 m, July 20; Kara-Dzhilga river, 4000–4500 m, July 22; Tas-Pestylk area, 4000–5000 m, July 25; Goo-Dzhiro river, 4500–5500 m, July 27; Shorluk river gorge, 4000–5000 m, July 28, 1942, Serp.; west. fringe of Tashkurgan town, along irrigation ditch in short grass meadow, June 13; same site, on debris cone, among boulders, June 13, 1959, Yun. and I-f. Yuan').

General distribution: Jung.-Tarb., Nor. and Cen. Tien Shan, East. Pam.; Mid. Asia (Alay).

Notes. 1. In coastal meadows, leaves measure up to 20 cm long and fruits 16 mm on stalk up to 10 mm long while inflorescence elongates in fruit.

2. A comparison of types of *O. globiflora* Bge. and *O. pagobia* Bge., like an analysis of their descriptions, did not reveal any differences. A study of the fairly large herbarium material for these 2, although different, species including the material from the classic localities also led to the same conclusion. It must be acknowledged that A. Bunge described the same species on different occasions, placing it in different sections—*Protoxytropis* Bge. and *Ianthina* Bge.

4. **O. latialata** P.C. Li in Acta Phytotax. sin. 18, 3 (1980) 370; id. in Fl. Xizang. 2 (1985) 857, tab. 282, fig. 8–14, descr. sin.

Described from Tibet (Chang Tang). Type in Beijing (PE).

On mountain slopes.

IIIB. Tibet: *Chang-Tang* (Bange—"Baingoin, alt. 5100 m, on slope, June 22, 1976—Qinghai-Xizang Compl. Exped. No. 9484, typus!").

General distribution: endemic ?

5. **O. lehmannii** Bge. in Arb. Naturf. Ver. Riga I (1847) 225; ej. Reliq. Lehmann. (1852) 76; ej. Sp. Oxytr. (1874) 10; B. Fedtsch. in Fl. Tadzh. 5 (1937) 515; Vassilcz. and B. Fedtsch. in Fl. SSSR, 13 (1948) 8; Fl. Tadzh. 5 (1978) 432; Filim. in Opred. rast. Sr. Azii [Key to Plants of Mid. Asia] 7 (1983); Fl. Xizang. 2 (1985) 856. —*O. aequipetala* Bge. in Mem. Ac. Sci. St.-Petersb. 22, 1 (1874) II. —**Ic.:** Fl. Uzbek. 3, Plate 71, fig. 2.

Described from Mid. Asia (Zeravshan basin, Kara Tau mountain range). Type in Paris (P).

On rocky and rubbly slopes, among rocks, on debris cones of brooks and gorges in upper mountain belt.

IIIB. Tibet: *South.* (Ali: "Pulan', 4000 m"—Fl. Xizang. l.c.).

IIIC. Pamir ?

General distribution: East. Pam.; Mid. Asia (Pam.-Alay).

Note. The correctness of its report for Tibet cannot be confirmed as Fl. Xizang. does not provide an illustration of this species.

6. **O. parasericopetala** P.C. Li in Acta Phytotax. sin. 18, 3 (1980) 369; id. in Fl. Xizang. 2 (1985) 855, tab. 281, fig. 1–7, descr. sin.

Described from Tibet (South.). Type in Beijing (PE).

In pastures ?

IIIB. Tibet: *South.* (Lhasa—"Lhasa, alt, 5000 m, in grassland, Aug. 18, 1965—Y.-t. Chang, K.-y. Lang, No. 1877, typus!").

General distribution: endemic ?

7. **O. pauciflora** Bge. Reliq. Lehmann. (1852) 77 in clavi, ej. Sp. Oxytr. (1874) 15; Saposhn. Mong. Altay (1911) 363; Kryl. Fl. Zap. Sib. 7 (1933) 1722; B. Fedtsch. and Vassilcz. in Fl. SSSR, 13 (1948) 15; Grub. Konsp. Fl. MNR (1955) 192; Fl. Kazakhst. 5 (1961) 341; Hanelt and Davazamc in Feddes Repert. 70 (1965) 42; Ulzij. in Issl. fl. i rast. MNR [Study of Flora and Plants of Mongolian People's Republic] I (1979) 112; id. in Grub.

Opred. rast. Mong. [Key to Plants of Mongolia] (1982) 165; Opred. rast. Tuv. ASSR [Key to Plants of Tuva Autonomous Soviet Socialist Republic] (1984) 152; C.Y. Yang in Claves pl. Xinjiang. 3 (1985) 89; Fl. Xizang. 2 (1985) 852. —Ic.: Fl. Xizang. 2, tab. 280, fig. 15–21.

Described from West. Siberia (Altay). Type in St.- Petersburg (LE).

On rocky slopes, rocks, rocky placers and talus and along pebble beds of rivers in alpine belt.

IA. Mongolia: *Khobdo* ("Oigur, Tsagan-Kobu"—Sap. l.c.), *Mong. Alt.* ("Bzau-Kul'"—Sap. l.c.), *Gobi-Alt.* (Dzun-Saikhan mountain range, west. part, steppe slope in upper belt, June 9; same site, nor. part, upper third of nor. slope, on bank, June 9, 1945, Yun.; Ikhe-Bogdo mountain range, nor. slope, Bityuten-Ama, midportion, on pebble beds near water, July 7, 1945—Yun.; "Nord Teil des Ich-Bogd, kleine Rasen an eimem Wasserlauf eines Hochtales, 1800 m, No. 2150", June 1962—Hanelt and Davazamc l.c.).

IIA. Junggar: *Tien Shan* (high plateau of Malyi Yuldus, on rocks, June 13, 1877—Przew.; C.Y. Yang l.c.).

IIIB. Tibet: (Fl. Xizang. l.c.), *Weitzan* (alpine belt along Konchyunchyu river, 3950–4250 m, June 3, 1884 3, 1884—Przew.).

General distribution: West. Sib. (Altay), East. sib. (Sayans).

8. **O. platysema** Schrenk in Bull. Ac. Sci. St.-Petersb. 10 (1841) 254; id. in Fischer et Meyer, Enum. pl. nov. Schrenk lect. 2 (1842) 47; Bge. Sp. Oxytr. (1874) 18; B. Fedtsch. and Vassilcz. in Fl. SSSR, 13 (1948) 15; Fl. Kirgiz. 7 (1957) 365; Fl. Kazakhst. 5 (1961) 340; Ikonnik. Opred. rast. Pamira [Key to Plants of Pamir] (1963) 172 ?; Filim. in Opred. rast. Sr. Azii [Key to Plants of Mid. Asia] 7 (1983) 339; C.Y. Yang in Claves pl. Xinjiang. 3 (1985) 89; Fl. Xizang. 2 (1985) 582. —Ic.: Fl. Kazakhst. 5, Plate 48, fig. 6; Claves pl. Xinjiang. 3, tab. 6, fig. 1.

Described from East. Kazakhstan (Jung. Ala Tau). Type in St.-Petersburg (LE).

On rock screes, moraines, small coastal meadows, meadow slopes and small grasslands in alpine belt.

IIA. Junggar: *Tarb.* (Saur mountain range, upper Bol'shoi Ob river, rocky tundra, June 13, 1904—Sap.), *Tien Shan* (nor. slope of Irenkhabirg, Dumbedan—Kum-daban, 2450–2750 m, May 28; Kumbel', 2750–3050 m, May 30; Kumbel', May 31; Kumbel', 3050 m, June 3; Bagaduslung, west. tributary of Dzhin, 2150–2750 m, June 4; Borgaty pass, 2450–2750 m, June 7; Naringol near Tsagan-Usu, branch of Dzhin, June 10; Chungur-daban pass, north of Borborogusun, 3050 m, June 13; Aryslyn, 2750–3050 m, July; valley of Kash river, south. flank, 2750 m, Aug. 17, 1879, A. Reg.; Urten-Muzart, Aug. 2, 1879—Fet.; around Mukhurdai pass, 2450–2750 m, July 19, 1893—Rob.; Ishigart pass from Iirtash to Akshiirak, tundra and rocky placers, June 30, 1902—Saposhn.; left bank of Manas river basin, upper Ulan-Usu valley along road from Danu pass, meadow on moraine, July 18; same site, valley of Danu river near Se-Daban pass, turf-covered placers, July 21, 1957, Yun. and I.-f. Yuan').

IIIA. Qinghai: *Nanshan* ([near Machan-Ula mountain], on stony placer, 3350–3650 m, July 23, 1879—Przew.; nor. slope of Humboldt mountain range, 3350–3650 m, June 24; upper Sulekhe river, 3950 m, June 25; mountains and terrain feature of Yamatyn-Umru, 3650 m, July 21, 1894, Rob.).

IIIB. Tibet: *Weitzan* (Burkhan-Budda mountain range, nor. slope, Nomokhun gorge, 3950–4250 m, on rocky floor, May 21–22, 1900; Amnen-Kor mountain range, south. slope, 4100 m, along meadowy mountain slopes, June 9, 1900; Yangtze river Basin, Shurchyu and Varmunchyu rivulets, 4250 m, on meadowy banks, May 19, 1901—Lad.).

IIIC. Pamir (around Ucha pass, June 17, 1909—Divn.; Kokat pass region, 3800–4000 m, June 16, 1942—Serp.).

General distribution: Jung.-Tarb., Nor. and Cen. Tien Shan, East. Pam.; Mid. Asia (Alay mountain range).

9. **O. proboscidea** Bge. in Mem. Ac. Sci St.-Petersb. 7, 22 I (1874) 17; Pamp. Fl. Carac. (1930) 151. —? *O.* sp. near *O. tatarica* Jacquem. Alcock. Rep. Natur. Hist. Results Pamir Boundary Commiss. (1898) 21.

Described from Kashmir (W. Tibet). Type in London (K) ?

On arid rocky mountain slopes.

IIA. Kashgar: *West.* (upper Kizylsu river beyond Kashgar between Simkhan to Egin, arid rocky slopes, July 1, 1929—Pop.)

IIIC. Pamir: ("Pamir region, 3950–4250 m, No. 17701"—Alcock l.c.).

General distribution: Himalayas (Kashmir).

24 10. **O. savellanica** Bge. ex Boiss. Fl. or. 2 (1872) 509; Bge. Sp. Oxytr. (1874) 15; B. Fedtsch. and Vassilcz. in Fl. SSSR, 13 (1948) 14; Fl. Kazakhst. 5 (1961) 341; Ikonnik. Opred. rast. Pamira [Key to Plants of Pamir] (1963) 174; Fl. Tadzh. 5 (1978) 437; Filim. in Opred. rast. Sr. Azii [Key to Plants of Mid. Asia] 7 (1984) 338; Fl. Xizang. 2 (1985) 856. —**Ic.:** Fl. Tadzh. 5, Plate 58, fig. 13–18.

Described from Fore Asia (nor. Iran). Type in Paris (P).

On rocky and rubbly slopes of alpine deserts and wastelands, 3000–5000 m alt.

IIIB. Tibet: *Chang Tang* (upper left bank tributaries of Tiznaf river 10 km before Seryk-Daban pass, alpine winter fat desert, 4600–4700 m, June 1; waterdivide of Tiznaf and Raskem, 217th km on Tibet highway 1–2 km east of Seryk-Daban Pass, 4780 m, winter fat desert, June 4, 1959—Yun. and I-f. Yuan'); *South* ("Shuankhe"—Fl. Xizang. l.c.).

General distribution: East. Pam.; Fore Asia, Caucasus (south.), Mid. Asia (West. Tien Shan, Pam.-Alay).

11. **O. sericopetala** C.E.C. Fischer in Kew Bull. Misc. Inform. 2 (1937) 95; Fl. Xizang. 2 (1985) 853. —**Ic.:** Fl. Xizang. 2, tab. 281, fig. 8–14.

Described from Tibet. Type in London (K).

IIIB. Tibet: *South.* ("Shigatse, 12,800 ft; Sept. 1935, No. 91—Cutting and Vernay—typus!; Lhasa, 11,500 ft, L.A. Waddell; Gyantse Hill, H.M. Stewart"; Gyantse, July–Sept. 1904, H.J. Walton).

General distribution: endemic.

12. **O. tatarica** Cambess. ex Bge. in Mem. Ac. Sci St.-Petersb. 7 ser. 22 I (1874) 16 excl. pl. kokanica; Henderson and Hume, Lahore to Jarkand (1873) 318; Baker in Hook. f. Fl. Brit. India, 2 (1876) 138; Hemsley in J.

Linn. Soc. Bot. (London) 30 (1894) 112; Deasy, In Tibet and Chin. Turk. (1901) 401; Hemsley, Fl. Tibet (1902) 174; Paulsen in Hedin, S. Tibet 6, 3 (1922) 62; Pamp. Fl. Carac. (1930) 151; Fl. Xizang. 2 (1985) 855. Described from Kashmir. Type in Paris (P).

In coastal meadows and sandy-pebbly banks of brooks, meadow slopes of mountains, rocks and rock screes, 4000–5000 m alt.

IIIB. Tibet: *Weitzan* (left bank of Yangtze river [near Konchyunchyu estuary], 4000 m, June 28, 1884-Przew.; "Toktomai Muren, 4250—4575 m, 1892, Rockhill"—Hemsley, 1902 l.c.), *Chang Tang* (Kuenluen, Province Khotan, Sumgal [on southern foot of Bushia Pass] to Gulbagashen [large Gashem quarries], Aug. 28–30, 1856—Schlagintweit; Tokuz-Daban mountain range, Muzlyk-Atasy mountain, 3950 m, on mountain slopes, Aug. 16; Przewalsky mountain range, nor. slope, 4250–4575 m, rocks and meadows, Aug. 23, 1890 Rob.; Pied ouest du Kinla, 5500 m, July 22; 5270 m July 7, 1892—Rhins; "Tibet Tsaidam, sandy soil in wide valley at 5350 m, 1891—Thorold; Goring valley, 90°25', 30°12', about 5050 m, 1895—Littledale; 88°20', 35°14', 4950 m, July 26, 1896—Wellby and Malcolm; 25 miles east of Lanak-La, 5200 m, 1896—Deasy and Pike"—Hemsley, 1902 l.c.; "Sarok-Tuz valley, 3950 m, 1898; Kizil-Chup, 4750 m, 1898"—Deasy l.c.; " Northern Tibet, Karyanak-Sai, Camp X, 3984 m, July 20; Jarkanlik, Chimen-Tagh, 3998 m, July 22, 1900, Hedin"—Paulsen l.c.); *South* (Ali: "Chzhada, 4200–4600 m"—Fl. Xizang. l.c.).

General distribution: Himalayas (Kashmir).

Section 2. Ianthina Bge.

13. **O. brevipedunculata** P.C. Li in Acta Phytotax. sin. 18, 3 (1980) 370; Fl. Xizang. 2 (1985) 860 .—**Ic.**: Fl. Xizang. 2, tab. 283, fig. 15–21. Described from Tibet. Type in Beijing (PE).

In alpine riparian meadows.

IIIB. Tibet: *Chang Tang* ("Shuang-Hu [Shuankhu], alt. 5200–5400 m, on meadow near the river, July 6, 1976, No. 11869, Qinghai-Xizang. Compl. Exp.—typus!").

General distribution: endemic ?

14. **O. chantengriensis** Vass. in Not. Syst. (Leningrad) 20 (1960) 231; Vassilcz. in Fl. SSSR, 13 (1948) 29, descr. ross.; Fl. Kirgiz. 7 (1957) 371; Filim. in Opred. rast. Sr. Azii [Key to Plants of Mid. Asia] 7 (1983) 342; Claves pl. Xinjiang. 3 (1985) 82.

Described from Cen. Tien Shan (Khan-Tengri). Type in St.-Petersburg (LE).

On rocky and rubbly desert slopes and trails and pebble beds of rivers in alpine belt.

IIA. Junggar: *Tien Shan* ("Temira gorge"—Claves pl. Xinjiang. l.c.).

General distribution: Nor. (Kungei-Ala Tau) and Cen. Tien Shan.

15. **O. coerulea** (Pall.) DC. Astrag. (1802) 68; Ledeb. Fl. Ross. I (1843) 589; Turcz. Fl. baic-dahur. I (1843) 304; Bge. Sp. Oxytr. (1874) 35; Peter-Stib. in Acta Horti Gotob. 12 (1937) 75; Kitag. Lin. Fl. Mansh. (1939) 291;

B. Fedtsch. and Vassilcz in Fl. SSSR, 13 (1948) 22; Boriss. in Fl. Zabaik. [Flora of Transbaikalia] 6 (1954) 599; Grub. Konsp. fl. MNR (1955) 189; Fl. Intramong. 3 (1977) 230; Fl. Tsentr. Sib. [Flora of Cen. Siberia] 2 (1979) 613; Ulzij. in Issl. fl. i rast. MNR [Study of Flora and Plants of Mongolian People's Republic] I (1979) 110; id. in Grub. Opred. rast. Mong. [Key to Plants of Mongolia] (1982) 164. —*O. subfalcata* Hance in J. Linn. Soc. Bot. (London), 13 (1872) 78; Franch. Pl. David. I (1884) 88; Forbes and Hemsl. Index Fl. Sin. I (1887) 167. —*Astragalus coeruleus* Pall. Reise, 3 (1776) 293 in nota. —*A. baicalensis* Pall. Spec. Astrag. (1800) 64. —Ic.: Pall. l.c. (1800) tab. 42, sub nom. *A. baicalensis*; Fl. Zabaik. [Flora of Transbaikalia] 6, fig. 305; Fl. Intramong. 3, tab. 115, fig. 9–16.

Described from East. Siberia (Transbaikal). Type in London (BM) ?

On steppe and meadow slopes, forest glades and borders, pebble beds along river valleys.

IA. Mongolia: *East. Mong.* (Barun-Sul mountain, nor. slope, upper part, cereal-grass steppe, July 1, 1971—Dashnyam, Karam. et al.; "Shilingol Ajmaq—[administrative territorial unit in Mongolia] Datsin'-Shan'"—Fl. Intramong. l.c.).

General distribution: *East. Sib.* (Angaro-Sayan, Daur.) Nor. Mong. (Hent., Mong.-Daur., Fore-Hing.), China (Dunbei, North, North-West).

16. **O. densa** Benth. ex Bge. in Mem. Ac. Sci. St.-Petersb. 7 ser. 22 (1874) 24; Baker in Hook . f. Fl. Brit. India, 2 (1876) 138; Hemsley, Fl. Tibet (1902) 173; Pamp. Fl. Carac. (1930) 151; Fl. Xizang. 2 (1985) 866. —*O. stipulosa* Kom. in Feddes Repert. 13 (1914) 231. —Ic.: Fl. Xizang. 2, tab. 286, fig. 8–14.

Described from Himalayas (Ladakh). Type in London (K). Map 4.

In alpine meadows, meadow slopes, sandy-pebble bed and stony banks and shoals of rivulets in alpine belt, 2500–5300 m alt.

IIA. Junggar: *Tien Shan* (Alta planities Malyi Julduss, 2300–2750 m, May 25 [July 6] 1877—Prz.; Oberer Taldy, 2400 m, May 17; Taldy, West Pass, 3350 m, May 20; Taldy quelle, Irenchabirga, 3050 m, May 21; Dumbedan—Kum-Daban, 2450–2750 m, Nordabhong des Irenchabirga, May 28; Kum-Daban, 2750 m, May 29—1979; Pass Baibeschan, zwischen Sarybulak und Kokkamyr, 2150–2750 m, Aug. 31, 1880—A. Reg.).

IIIA. Qinghai: *Nanshan* (Nanshan alps [south. slope of Humboldt mountain range near Machan-Ul mountain], 3350–3650 m, July 12 [24]; same site, 3350 m, July 14 [26] 1879—Przew.; Humboldt mountain range, nor. slope, along Kuku-Usu river, 2750–3350 m, on mounds, May 26; same site, Ulan-Daban area, 3650 m, June 21; same site, along Ulan-Bulak gorge, 3650 m, June 23; Ritter mountain range, east. continuation, Ikhe-Daban, 3650 m, June 27—1894, Rob.), *Amdo* (ad fl. Baga-Gorgi, 12,750 m, May 12 [24] 1880—Przew.).

IIIB. Tibet: *Chang Tang* (nor. slope of Russky mountain range along Moldzha river, 2200 m, on pebble bed, May 4 [16] 1885—Przew., typus *O. stipulosa* Kom.; nor. slope of Russky mountain range, Aksu river, Bash-Bulak area, 3650 m, July 1; Przewalsky mountain range, nor. slope, 4250–4550 m, Aug. 23—1890, Rob.; [in Aru-Cho lake region] 17,500 ft [5300 m] camp 11, 1891, Thorold"—Hemsley l.c.; vallee Koutaslik, 5300 m, Aug. 1892—Rhins; "Ban'ge, Zhitu, Shuankhu"—Fl. Xizang. l.c.); *Weitzan* (ad ripam fl. Bytschu, May 3 [15]; fauce ad rivulum prope cacumine jugi inter Hoangho et Jangtze, May 30 [June 11]; ad fl. Talatschu, June 6 [18]; ad fl. Dshagyn Gol, 4200 m, July 11 [23]; ad fl. Ladronum, 4100 m, July 14 [26]—1884, Przew.; Burkhan-Budda mountain range, south. slope, 3950–4250 m, June 2; Russkoe lake and bank of Huang He, 4100 m, June 17, 1900; Amnen-Kor mountain range, nor. slope, 4100 m, July 6, 1901—Lad.).

General distribution: Himalayas (Kashmir).

17. O. dumbedanica Grub. et Vass. in Novit. Syst. Pl. Vasc. 24 (1987) 134.

Described from Sinkiang (Tien Shan). Type in St.-Petersburg (LE).

In montane meadows and meadow and steppe slopes, 2400–3100 m alt.

IIA. Junggar: *Tien Shan* (Upper Taldy, 2150–2450 m, May 16; same site, 2750–3050 m, May 20; Dumbedan-Kumdaban, nor. slope of Irenkhabirga mountain range, 2450–2750 m, May 22; Taldy, 2450–2750 m, May 27; Kumbel', 2750–3050 m, May 31—1879, A. Reg., typus!; Bagaduslung, west tributary of Dzhin, 2150–2750 m, June 4, 1879—A. Reg.; along road to Shipaotszy near Turfan from Sansanko, 2400 m, steppe, No. 5667, June 16, 1958—Lee and Chu (A.R. Lee (1959)).

General distribution: endemic.

18. O. filiformis DC. Astrag. (1802) 80 excl. syn.; Ledeb. Fl. Ross. I (1843) 590; Turcz. Fl. baic.-dahur. I (1843) 305; Bge. Sp. Oxytr. (1874) 34; Pavl. in Byull. Mosk. obshch. isp. prir., otd. biol. 38 (1929) 93; Kitag. Lin. Fl. Mansh. (1939) 291; B. Fedtsch. and Vassilcz. in Fl. SSSR, 13 (1948) 24; Boriss. in Fl. Zabaik. [Flora of Transbaikalia] 6 (1954) 600; Grub. Konsp. fl. MNR (1955) 189; Hanelt et Davazamc in Feddes Repert. 70 (1965) 42; Fl. Intramong. 3 (1977) 227; Fl. Tsentr. Sib. [Flora of Cen. Siberia] 2 (1979) 614; Ulzij. in Issled fl. i rast. MNR [Study of Flora and Vegetation of Mongolian People's Republic] I (1979) 111, id. in Grub. Opred. rast. Mong. [Key to Plants of Mongolia] (1982) 164; Opred. rast Tuv. ASSR [Key to Plants of Tuva Autonomous Soviet Socialist Republic] (1984) 152. —Ic.: Fl. Intramong. 3, tab. 116, fig. 1–6; DC. l.c. tab. 4.

Described from East. Siberia (Dauria). Type in Geneva (G).

On steppe rubbly and rocky slopes of mountains and valleys, on slopes and crests of ridges and mud cones, pebble beds and rocks on banks.

IA. Mongolia: *Mong. Alt.* (Khan-Taishri-Ula, 2600-3700 m, outlier rocks, Aug. 10, 1945—Leont'ev), *Cen. Khalkha* (Gagtsa-Khuduk, June 23, 1851—Tatarinov; near Ugei-Nor lake, July 14, 1924—Pavl.; Kharukhe river source, Aug. 12 and 28; Ubur-Dzhargalante river between source and Agit mountain, Aug. 25—1925, Krasch. and Zam.; vicinity of Ikhe-Tukhum-Nor lake, Nuramta mountain, June; Ikhe-Tukhum-Nor lake—Mishim-Gun, June 1926—Zam.; vicinity of Kholt area near Sondzhi, July 16–17, 1926—Gus.; mountain range 15 km nor.-east of Santu somon, Aug. 24, 1951—Kal.; Bayan-Dzurkh somon, Kholtyn-Daba pass, July 2, 1970—Banzragch, Karam. et al.; Munkh-Khan-Ula, 1607 m, July 25, 1974—Golubkova and Tsogt), *East. Mong.* (in locis subarenosis Mongolia chinensis australis, 1831—I. Kuznetsov; kuitun, May 19; Borol'dzhi [Dagkhan hills], July 17, 1841—Kirilov; Mongolia chinensis, Scharty, May 21—July 2, 1850—Tatarinov; prope Gurbunei-Bulak inter Kulussutajevsk et Dolon-Nor, 1870—A. Lomonossov; inter Kalgan et mong. Inschan, May-June, 1871; mont.

Sumahada, 1871—Prz.; between Dabastu-Nor and Mandybai lake, May 30; Elisyn-Khuduk well, June 5; Ulan-Dzhilgu mountain range, July 3; Kholyn-Gol river valley, July 22—1899, Pot. and Sold.; Khailar railway station, June 10, 1902—Litw.; around Manchuria station, 1915—Nechaeva; 40 km nor. of Khailar station, beyond Mergel brook, Aug. 19, 1925—I. Kozlov; Manchuria station, Nanshan mountain, 650 m, slope, No. 880, June 24, 1951—Wang Chang et al.; Tsagan-Obo somon, Tsagan-Chulut, nor. slope of hills, Aug. 3, 1956—Dashnyam; Datsin'shan' mountain range 20 km nor. of Khukh-Khoto town, near pass, June 4, 1958—Petr.; Shilin-Khoto town, steppe, 1959—Ivan.; 26 km south-east of Underkhan town, Bayan-Khuduk hills, summit, 1580 m, June 19; Barun-Sul hill, upper part of nor. slope, July 1, 1971, Dashnyam, Isachenko et al.; Shiliin-Bogdo mountain, crest, July 8, 1971—Dashnyam, Karam. et al.), *Depr. Lakes* (on southern descent of Tannu-ol around Ak-Karasuk river, on southern slopes, July 6, 1892—P. Krylov), *Val. Lakes* (along Ologoi river, July 10; along Uta river, July 11, 1924—Gorbunova; Gun-Narin somon, 4–5 km south of Khundulengiin-Nur lakelet, hillocky plain, July 27, 1952—Davazamc), *Gobi-Alt.* (Baga-Bogdo, canyon terraces, alt. 6000–8500 ft, No. 264, 1925—R. Chaney).

General distribution: *East. Sib.* (Sayan., Daur.) Nor. Mong.

19. **O. humifusa** Kar. et Kir. in Bull. Soc. natur. Moscou, 15 (1842) 535; Bge. Sp. Oxytr. (1874) 28; Vassilcz. and B. Fedtsch. in Fl. SSSR, 13 (1948) 29; Fl. Kirgiz. 7 (1957) 371; Fl. Kazakhst. 5 (1961) 348; Ikonnik. Opred. rast. Pamira [Key to Plants of Pamir] (1963) 21; Filim. in Opred. rast. Sr. Azii [Key to Plants of Mid. Asia] 7 (1983) 341; Claves pl. Xinjiang. 3 (1985) 83; Fl. Xizang. 2 (1985) 861. —*O. immersa* (Baker ex Aitch.) Bge. ex B. Fedtsch. in Beih. Blt. Centralbl. 22 (1907) 212; Vassilcz. and B. Fedtsch. in Fl. SSSR, 13 (1948) 25; Fl. Kirgiz. 7 (1957) 369; Fl. Kazakhst. 5 (1961) 345; Ikonnik. Opred. rast. Pamira [Key to Plants of Pamir] (1963) 174; Filim. in Opred. rast. Sr. Azii [Key to Plants of Mid. Asia] 7 (1983) 340. —*O. incanescens* Freyn in Bull. Herb. Boiss. 2 ser., 5 (1905) 1023; Filim. in Opred. rast. Sr. Azii [Key to Plants of Mid. Asia] 7 (1983) 340. —*O. pamirica* Danguy in Bull. Mus. nat. hist. natur. 14 (1908) 130. —*Astragalus immersus* Baker ex Aitch. in J. Linn. Soc. (London) Bot. 18 (1881) 45. —**Ic.:** Fl. Tadzh. 5, Plate 9, fig. 1–7, 15–21.

Described from East. Kazakhstan (Jung. Ala Tau). Type in St.-Petersburg (LE).

On rubbly and rocky slopes, moraines, rocks and talus, small alpine meadows and pebble beds of rivers, mossy tundra in upper mountain belt, 3000–5000 m alt.

IB. Kashgar: *West.* (Sarykol mountain range, Bostan-Terek area, July 11, 1929—Pop.; Torugart settlement, 3600 m, June 20, 1959—Yun. and I-f. Yuan').

IIA. Junggar: *Tarb.* ("Dachen"—Claves pl. Xinjiang. l.c.), *Tien Shan* (ibid l.c.).

IIIB. Tibet: *Chang Tang* (upper Tiznaf river 15 km beyond kyude settlement on Tibet highway, June 1, 1959—Yun. and I.-f. Yuan'; "west. Tibet"—Fl. Xizang. l.c.), *Weitzan* (Burkhan-Budda mountain range, Nomokhun river gorge, 3650–3950 m, May 20, 1900—Lad.; "Sosyan'"—Fl. Xizang. l.c.), *South.* ("Dansyun"—ibid l.c.).

IIIC. Pamir ("Mouztag-Ata, Bour-Teppe, 3380 m, July 23, 1906, Lacoste"—Danguy l.c.; Billuli river at its confluence into Chumbus river, June 11; around Ucha pass, June 17; Kok-Muinak pass, July 27—1909, Divn.; Karatash area, 3500–4000 m,

Aug.-Sept. 1941; Pil'nen gorge, 3000–4000 m, June 30; same site, 4500–5000 m, tundra, July 14; moraine-covered waterdivide between Atrakyr and Tyuzutek rivers, mossy tundra, 4500–5000 m, July 20; Taspestlyk area, 4000–5000 m, July 25; Tenboin pass, 4200–4300 m, Aug.-Sept.—1942, Serp.).

General distribution: Jung.-Tarb., Nor. and Cen. Tien Shan, East. Pam.; Fore Asia (east.), Mid. Asia (West. Tien Shan, Pam.-Alay).

Note. A comparison of authentic specimens preserved in St.-Petersburg—isotype of *O. immersa* and type with dozens of isotypes of *O. humifusa*—did not reveal any significant differences between them. The same conclusion reached after analysing the fairly large herbarium material whose identification led to surprising differences, with some specialists identifying the same specimens as *O. immersa* and others as *O. humifusa*. Local "Flora" and "Keys" point out that mature fruits of *O. humifusa* are pendent and stipules sparsely pilose while mature pods of *O. immersa* are erect and stipules glabrous. It is not so in reality: mature pods in all cases are pendent while stipules of younger leaves are pilose to some extent but soon becoming glabrescent. Pendent pods and pilose stipules may also be seen in the above-cited isotype of *O. immersa*. The length of pods varies from 10 to 20 mm. Similarly, the sizes of all plant parts and extent of pubescence vary greatly over the vast distribution range of this species relative to the habitat conditions.

20. **O. imbricata** Kom. in Feddes Repert. 13 (1914) 232; Peter-Stib. in Acta Horti Gotob. 12 (1937) 76; Hao in Bot. Jahrb. 68 (1938) 614; Walker in Contribs. U.S. Nat. Herb. 28 (1941) 638. —*O. holanshanensis* H.C. Fu in Acta phytotax. sin. 20, 3 (1982) 113; Pl. vasc. Helansh. (1986) 157. —*O. kanitzii* Simps. in Notes Roy. Bot. Gard. Edinb. 8 (1915) 263. —*Astragalus loczii* Kanitz var. *scaposa* Kanitz, A novenytani... (1891) 17; id. Wissensch. Ergebn. 2 (1898) 693. —Ic.: Pl. vasc. Helansh. tab. 28.

Described from Nor.-West China (south. kansu: Dzhoni). Type in St.-Petersburg (LE). Map 3.

On Steppe sandy-pebbly and rocky slopes, rocks and precipices, pebble beds of rivers, 1800–3900 m alt.

28 **IA. Mongolia:** *Alash. Gobi* (Alashan mountain range, Dartymto gorge, south. slope, midbelt, May 23, 1908—Czet.; "Helanshan, July 26, 1962, Y.C. Ye 321, typus *O. holanshanensis*!; Helan Shan Nansi, ad radices montium alt. 2030 m, Aug. 2, 1963, Y.C. Ma 242; Helan Shan, Daliumengou, alt. 2400 m, Aug. 31, 1963, Y.C. Ma 228"— H.C. Fu l.c.), *Khesi* (Loukhushan' mountain range, south. slope, middle and lower belts, July 17, 1908—Czet.).

IIIA. Qinghai: *Nanshan* (Sharagol'dzhin river, Sunginor area, about 3000 m, July 9; same site, Paidza-Tologoi area, July 11, 1894—Rob.; nor. Slope of Humboldt mountain range, Chonsai area, 3050–3350 m, July 23, 1895—Rob.; Xining mountains, Myndan'sha river, July 1, 1890—Gr.-Grzh.; "inter Lantschou, Pingfan et Gaogai, No. 2182, Fenzl"—Peter-Stib. l.c.; "in der Nahe des Klosters Taschinsze, um 3900 m"— Hao l.c.; "Hsin-Cheng, west of Lanchow, 1923, No. 309, R.C. Ching"—Walker l.c.; high Nanshan foothills 60 km south-east of Chzhan'e town, 2200 m, July 12; Altyntag mountain range 15 km south of Aksai settlement, 2800 m, Aug. 2, 1958; pass through Altyntag 24 km from Aksai settlement along road to Qaidam, 3460 m, Aug. 2; 33 km west of Xining town, 2450 m, Aug. 5; 66 km west of Xining town, 2800 m, Aug. 5— 1959, Petr.), *Amdo* (ad fl. Hoangho prope ostium fl. Tschurmyn, 2600–2750 m, May 8 [20]; ad affl. fl. Baga Gorgi, April 27 [May 9]; ad fl. Hoangho super. in ripa clata glareosa, 2450–2750 m, May 19 [31]; prope alpem Dschachar-Dschargyn, 3200–3500 m, June 9 [21]—1880, Przew.; San'chuan', on floor of basin, 1800 m, April 8; Ar'ku village near Karun river esturary on Huang He, May 5; Lanchzhulunva river valley, May 14; Urun'vu river valley, May 15—1885, Pot.; "Radja-Rock, No. 14004"—Peter-Stib. l.c.).

General distribution: China (Kansu south.).

Note. The species is vicarious to *O. merkensis* Bge. and very similar to it but more xeromorphic, densely pubescent, with characteristic columnar shoots, imbricately covered with overlapping stem-like bases of dead leaves with stipules. The colour of corolla varies from dun to blue and light-violet.

21. **O. krylovii** Schipz. in Not. Syst. (Leningrad) 1, 7 (1920) 1; Saposhn. in Izv. Tomsk. otd. Russk. bot. obshch. I (1921) 32; Kryl. Fl. Zap. Sib. 7 (1933) 1725; Vassilcz. and B. Fedtsch in Fl. SSSR, 13 (1948) 21; Fl. Kazakhst. 5 (1961) 344; Ulzij. in Issl. fl. i rastit. MNR [Study of Flora and Vegetation of Mongolian People's Republic] 1 (1979) 110; id. in Grub. Opred. rast. Mong. [Key to Plants of Mongolia] (1982) 165; Claves pl. Xinjiang. 3 (1985) 85. —Ic.: Fl. Kazakhst. 5, Plate 18, fig. 2; Fl. SSSR, 13, Plate 1, fig. 3.

Described from Altay (Narym mountain range). Type in St.-Petersburg (LE).

In solonetzic sedge meadows, rocky and rubbly slopes of mountains rocks in alpine belt.

IA. **Mongolia:** *Mong. Alt.* (Kobdo river basin, Khashyatu-Gol near Bayan-Enger somon, 2050 m, June 30; Duro-Nur lake, east. coast, Buratiin-Gol, 2400 m, June 30; Dayan-Nur lake, south. extremity, Yamatyn-Ula, peak. 2550 m, July 9, 1971, Grub., Ulzij. et al.).

General distribution: West Sib. (Altay).

22. **O. kumbelica** Grub. et Vass. in Novit. Syst. Pl. Vasc. 24 (1987) 135. Described from Sinkiang (Tien Shan). Type in St.-Petersburg (LE). In alpine meadows ?

IIA. **Junggar:** *Tien Shan* (Kumbel', 3050 m, June 3, 1879—A. Reg., typus!; in Turfan, 2500 m, on shaded slope, No. 5781, June 23, 1958—Lee and Chu).

General distribution: endemic.

23. **O. ladyginii** Kryl. in Acta Horti Petrop. 21 (1903) 5; Saposhn. Mong. Altay (1911) 363; Kryl. Fl. Zap. Sib. 7 (1933) 1724; Vassilcz. and B. Fedtsch. in Fl. SSSR, 13 (1948) 22; Grub. Konsp. fl. MNR (1955) 190; Fl. Kazakhst. 5 (1961) 345; Ulzij. in Issl. fl. i rastit. MNR [Study of Flora and Vegetation of Mongolian People's Republic] (1979) 110; id. in Grub. Opred. rast. Mong. [Key to Plants of Mongolia] (1982) 164; Claves pl. Xinjiang. 3 (1985) 84. — Ic.: Acta Horti Petrop. 21, Plate 3.

Described from Mongolia. Type in St. Petersburg (LE).

On steppe and desert-steppe rubbly and rocky slopes and trails of mountains and river valleys.

29   **IA. Mongolia:** *Khobd.* (Altyn-Khatasyn area, on border, dry sandy soil, June 17, 1879—Pot.; Oigur river [Oiguriin-Gol], rocky plain, July 29, 1899—Lad., typus!; " Kalgutty, alp."—Sap. l.c.).
General distribution: West. Sib. (Altay).

24. **O. larionovii** Grub. et Vass. in Novit. Syst. Pl. Vasc. 24 (1987) 136.
Described from Sinkiang (Tien Shan). Type in St.-Petersburg (LE).
On steppe slopes of mountains ?

IIA. Junggar: *Tien Shan* (middle course of Taldy, May 26, 1879—A. Reg.; typus!; Taldy river gorge [= Epte or Turgen'] in Irenkhabirga hills, 1200 m, May 27; Bagaduslung-Bainamun, 1850 m, June 4—1879, A. Reg.).
General distribution: endemic.

25. **O. linearibracteata** P.C. Li in Acta Phytotax. Sin. 18, 3 (1980) 371; Fl. Xizang. 2 (1985) 863. —Ic.: Fl. Xizang. 2, tab. 284, fig. 8–14.
Described from Tibet (Weitzan). Type in Beijing (PE).
On rocky slopes, moraines, debris cones in alpine belt.

IIIB. Tibet: Weitzan ("Sog-Xian [Sosyan'], alt. 4200 m, on alluvial cone, June 4, 1961, No. 3363, Wang Chin-ting-typus!"—P.C. Li l.c.).
General distribution: endemic.

26. **O. lutchensis** Franch. in Bull. Mus. hist. natur. 3 (1897) 322, descr. manca; Grub. in Novit. Syst. Pl. Vasc. 27 (1990) 93. —*O. atbaschi* Saposhn. in Not. Syst. (Leningrad) 4 (1923) 131; Vassilcz. and B. Fedtsch. in Fl. SSSR, 13 (1948) 24; Fl. Kirgiz. 7 (1957) 366; Fl. Kazakhst. 5 (1961) 345; Filim. in Opred. rast. Sr. Azii [Key to Plants of Mid. Asia] 7 (1983) 339.
Described from Tibet (Chang Tang). Type in Paris (P).
On rocky slopes, rock talus and moraines in upper mountain belt.

IB. Kashgar: *West.* (Sarykol mountain range west of Kashgar, Boston-Terek area, July 10, 1929—Pop.).
IIIB. Tibet: *Chang Tang* (Keriya mountain range, along Araldyk river, 2900 m, June 26 [July 8]; same site, along Khan-Yut river, 3650 m, July 13 [25]—1885, Przew.; Bassin de la Lutche (affluent du Keria Daria), No. 2, rec. le April 9, 1892—D. de Rhins, typus ! [P]).
General distribution: Nor. and Cen. Tien Shan, East. Pam.

Note. Franchet (op. cit.) does not provide either description or diagnosis of this species but only refers to its affinity: "*O. tilingii* valde affinis, sed breviter caulescens; stipulae pallidae, membranaceae, nec fusca, demum coriaceae; flores paulo minores, probabiliter lutescentes, nec purpurascentes". It is impossible to identify the species from the above remarks since far-eastern *O. tilingii* Bge. falls in section *Orobia* of subgenus *Eumorpha* while Franchet's species undoubtedly belongs to section *Ianthina* of subgenus *Oxytropis.*

27. **O. melanotricha** Bge. in Mem. Ac. Sci. St.-Petersb. 7 ser. 22 (1874) 26; Vassilcz. and B. Fedtsch. in Fl. SSSR, 13 (1948) 30; Fl. Kirgiz. 7 (1957) 372; Ikonnik. Opred. rast. Pamira [Key to Plants of Pamir] (1963) 172; Filim. in Opred. rast. Sr. Azii [Key to Plants of Mid. Asia] 7 (1983) 341; Claves pl. Xinjiang. 3 (1985) 83. —*O. humifusa* var. *grandiflora* Bge. in Mem. Ac. Sci. St.-Petersb. 7 ser. 14, 4 (1869) 44.

Described from Cen. Tien Shan (Tashrabat mountain range). Type in Paris (P).

On pebble beds of brooks and rivulets, moraines, talus and rubbly slopes in alpine belt.

IIA. Junggar: *Tien Shan* ("south. slope of Tien Shan"—Claves pl. Xinjiang. l.c.).
IIIC. Pamir (Claves pl. Xinjiang. l.c.).
General distribution: Cen. Tien Shan, East. Pam.; Mid. Asia (Pam.-Alay).

Note. Did not examine any specimen from Cen. Asia.

30   28. **O. merkensis** Bge. in Bull. Soc. natur. Moscou 39 (1866) 11; ej. Sp. Oxytr. (1874) 33; id. in Izv. obshch. lyubit. estestv. 26 (1880) 172; Hedin, S. Tibet, 6, 3 (1922) 63; B. Fedtsch. and Vassilcz. in Fl. SSSR, 13 (1948) 27; Fl. Kirgiz. 7 (1957) 370; Fl. Kazakhst. 5 (1961) 347; Filim. in Opred. rast. Sr. Azii [Key to Plants of Mid. Asia] 7 (1983) 342; Claves pl. Xinjiang. 3 (1985) 82. —*O. avis* Saposhn. in Not. Syst. (Leningrad) 4 (1923) 131; Vassilcz. and B. Fedtsch. in Fl. SSSR, 13 (1948) 26; Fl. Kazakhst. 5 (1961) 346; Filim. in Opred. rast. Sr. Azii [Key to Plants of Mid. Asia] 7 (1983) 342; Claves pl. Xinjiang. 3 (1985) 85. —*O. ervicarpa* Vved. ex Filim. in Not. Syst. (Taschkent) 20 (1982) 41; Filim. in Opred. rast. Sr. Azii [Key to Plants of Mid. Asia] 7 (1983) 342. —Ic.: Fl. SSSR, 13, Plate 5, fig. 1; Fl. Kazakhst. 5, Plate 49, fig. 1 and 3.

Described from Nor. Tien Shan. Type in St.-Petersburg (LE).

On rubbly and rocky steppe and turf-covered meadow slopes, pebble beds on banks, terraces and precipices, steppe to alpine belt (1500–3500 m).

IB. Kashgar: *Nor.* (Khanga river valley 25 km nor.-west of Balinte settlement along road to Yuldus from Karashar, steppe belt, Aug. 1; Muzart river valley 7–8 km beyond exit from gorge near Kurgan settlement along road to Oi-Terek area, steppe belt, Sep. 7—1958, Yun. and I-f. Yuan'; Kucha—Shakh'yar, 2290 m, No. 10076, July 7, 1959—Lee and Chu), *West.* ("Suukty valley around Kashgar"—Bunge, 1880, l.c.).
IIA. Junggar: *Tien Shan* (Urtas-Aksu, June 17; Sairam lake, July 19; Urtak-Sary west of Sairam lake, July 19; west of Sairam lake, July 20—1878 , Fet.; Khonakhai river, 1200 m, June 17; Chapchal pass, 1500–2100 m, June 28; Bogdo nor. of Kul'dzha, 1800–2100 m, July 24; Kyzemchek-Sairam, 2100–2750 m, July 29; Sairam-Nor, Aug. 1878; Bagaduslung-Bainamun, 1800 m, May 8; same site, June 4; upper Taldy, 1500 m, May 22; south. source of Taldy, 2450–3050 m, May 25; Kumbel', 2750–3050 m, May 31; Bainamun-Dzhin, 1500–1800 m, June 5; Tsagan-Tyunge, 1500–1800 m, June 8; Naryn-Gol near Tsagan-Usu, June 10; Borborogusun, 1800 m, June 15: same site, 2750 m, June 15; Aryslyn, 2450 m, July 18—1879, A. Reg.; Kukurtuk Tal, 3000–4000 m, end of June, 1903—Merzb.; Manas river basin, Ulan-Usu river valley at its confluence with Dzhartas, subalpine meadow, July 18, 1957—Yun. and I-f. Yuan'; 30 km west of Chzhaosu, Tszin'tsyuan' settlement, on river bank, No. 3265, Aug. 12, 1957—Kuan; Tekes river valley 1 km east of Kobo settlement along road to Kzyl-Kure, on precipice of terrace, Aug. 23, 1957—Yun. and I-f. Yuan'; B. Yuldus basin 3–4 km south-west of Bain-Bulak settlement, Khaidyk-Gol river, pebble bed terrace, Aug. 10; same site 1 km south-east of of Bain-Bulak settlement, Sept. 9; same site, 1–2 km east of Bain-Bulak settlement, steppe belt, Sept. 9—1958, Yun. and I-f. Yuan').

IIIC. *Pamir* ("eastern Pamir, Kara-jilga, valley and spring at Basik-kul, 3727 m, July 24, 1894"—Hedin l.c.).

General distribution: Jung.-Tarb. Nor. and Cen. Tien Shan; Mid. Asia (Pam.-Alay).

Note. 1. The corolla of this species shows considerable variation of colour tints—pure dun (or pale-yellow) and dun with violet keel to blue and violet. The large-fruited variety—var. *ervicarpa* (Vved.) Vass. l.c.—is found all over the distribution range of this species with no distinct area: large oblong-oval pods are seen on plants with dun-coloured flowers and, contrarily, small round pods on plants with violet flowers, there being several intermediates between the extreme forms of pods.

2. Apart from 4 authentic specimens of *O. avis* Saposhn., the Herbarium of the Komarov Botanical Institute of the Russian Academy of Sciences has no specimen under this name. The difference in the length of cusp between *O. avis* and *O. merkensis* cited in the key by Z. Filimonova (l.c.) is not noticed in nature; short, up to 1 mm long, cusps are found, frequently, among typical specimens of *O. merkensis* including those tested by A. Bunge himself. *O. avis* is simply a small-sized alpine form of *O. merkensis*.

29. **O. nutans** Bge. in Bull. Soc. natur. Moscou, 39, 2 (1866) 61; ej. Sp. Oxytr. (1874) 37; Vassilcz. and B. Fedtsch. in Fl. SSSR, 13 (1948) 37; Fl. Kirgiz. 7 (1957) 372; Fl. Kazakhst. 5 (1961) 350; Filim. in Opred. rast. Sr. Azii [Key to Plants of Mid. Asia] 7 (1983) 351; Claves pl. Xinjiang. 3 (1985) 81.

31   Described from East. Kazakhstan (Jung. Ala Tau, Balykty river). Type in St.-Petersburg (LE).

On rubbly and melkozem steppe slopes in middle and lower mountain belts.

IIA. Junggar: *Jung. Ala Tau, Tien Shan* (Claves pl. Xinjiang. l.c.).
General distribution: Jung.-Tarb., Nor. and Cen. Tien Shan.

Note. We did not find even a single specimen of this species from Sinkiang in the Herbarium of the Komarov Botanical Institute of the Russian Academy of Sciences. The report in Claves pl. Xinjiang. l.c. to its occurrence in Dachen town (Chuguchak) region is highly doubtful.

30. **O. penduliflora** Gontsch. in Not. Syst. (Leningrad) 8, 11 (1940) 186; Vassilcz. and B. Fedtsch. in Fl. SSSR, 13 (1948) 38; Fl. Kirgiz. 7 (1957) 372; Fl. Kazakhst. 5 (1961) 350; Filim. in Opred. rast. Sr. Azii [Key to Plants of Mid. Asia] 7 (1983) 343; Claves pl. Xinjiang. 3 (1985) 81. —Ic.: Fl. Kazakhst. 5, Plate 19, fig. 2.

Described from Cen. Tien Shan (Koksu river). Type in St.-Petersburg (LE).

In meadows and meadow slopes, along river valleys in upper mountain belt.

IIA. Junggar: *Tien Shan* (Aryslyn, 2750 m, July 11, 15 and 17; source of Kash river, 3050–3350 m, Aug. 12—1879, A. Reg.; Koksai river east of Sarytyur pass, 2450 m, July 22, 1893—Rob.; B. Yuldus basin, south. slope of Narat mountain range along road to Bain-Bulak from Dasht pass, Aug. 8, 1958—Yun and I-f. Yuan'; "Sin'yuan', about 3000 m"—Claves pl. Xinjiang. l.c.).

General distribution: Jung. Ala Tau, Nor. and Cen. Tien Shan.

31. **O. pusilla** Bge. in Mem. Ac. Sci. St.-Petersb. 7 ser. 22, I (1874) 27; Fl. Xizang. 2 (1985) 860 sub. auct. Fisch. —**Ic.**: Fl. Xizang 2, tab. 283, fig. 8–14. Described from Himalayas (Kashmir). Type in St.-Petersburg (LE).

In moist and coastal meadows, river shoals in high mountain areas, 3800–5100 m alt.

IIIB. **Tibet:** *South.* ("Chzhada, Pulan', Sage"—Fl. Xizang. l.c.).

General distribution: Himalayas (Kashmir).

Note. Apart from authentic specimens cited by A. Bunge, we did not find any others in the Herbarium of the Komarov Botanical Institute of the Russian Academy of Sciences or anywhere else. *O. coelestis* Abduss., judging from its description and illustration [Fl. Tadzh. 5 (1978) 631, Plate 61, fig. 11], is identical to it. The description and illustration of this species in Flora of Tibet (Fl. Xizang., 2, l.c.) wholly corresponds to the type although the text erroneously cites its author as Fischer instead of Bunge.

32. **O. rupifraga** Bge. in Bull. Soc. natur. Moscou 39 (1866) 8; ej. Sp. Oxytr. (1974) 24; Vassilcz. and B. Fedtsch. in Fl. SSSR, 13 (1948) 25; Fl. Kirgiz. 7 (1957) 369; Fl. Kazakhst. 5 (1961) 346 ?; Filim. in Opred. rast. Sr. Azii [Key to Plants of Mid. Asia] 7 (1983) 341; Claves pl. Xinjiang. 3 (1985) 85.

Described from Cen. Tien Shan (Sary-Dzhas river). Type in St.-Petersburg (LE).

On rocky slopes and moraines, pebble beds of rivers in upper mountain belt, 2500–3500 m alt.

IIA. **Junggar:** *Tien Shan* (upper Taldy, 2450 m, May 16 and 17; same site, 2750–3050 m, May 20; west. pass of Taldy, 3350 m, May 20; Taldy, 2450 m, May 24; Kumdaban, 2750 m, May 29; east. tributary of south. Khapchagai opposite Karashar, 2450 m, Sept. 16; Khatyn-Bogdo near Karashar, 2450 m, Sept. 14—1879, A. Reg.; Uital river gorge, 2450–3350 m, May 31, 1889—Rob.; subliches Kiukonik Tal beim Lager unter Tschon-Yailak Pass, June 15, 1908—Merzb.; "Ili-Baichen"—Claves pl. Xinjiang. l.c.).

General distribution: Nor. and Cen. Tien Shan.

33. **O. saposhnikovii** Kryl. in Acta Horti Petrop. 21 (1903) 4; Saposhn. Mong. Alt. (1911) 363; Kryl. Fl. Zap. Sib. 7 (1933) 1723; Vassilcz. and B. Fedtsch. in Fl. SSSR, 13 (1948) 21; Grub. Konsp. fl. MNR (1955) 193; Fl. Kazakhst. 5 (1961) 344; Ulzij. in Issl. fl. i rastit. MNR [Study of Flora and Vegetation of Mongolian People's Republic] 1 (1979) 109; id. in Grub. Opred. rast. Mong. [Key to Plants of Mongolia] (1982) 165; Opred. rast. Tuv. ASSR [Key to Plants of Tuva Autonomous Soviet Socialist Republic] (1984) 152; Claves pl. Xinjiang. 3 (1985) 84. —**Ic.**: Acta Horti Petrop. 21, Plate 2, fig. 2; Fl. Kazakhst. 5, Plate 48, fig. 4.

Described from Altay (upper Chui river). Type in St.-Petersburg (LE).

In sedge-*Cobresia* meadows, coastal meadows, moraines, tundra in alpine belt, 2800–3400 m alt.

IA. **Mongolia:** *Khobd.* ("Oigur"—Sap. l.c.: Koshagach steppe up to Kobdo upper courses, spring and summer, 1897—S. Demidova). *Mong. Alt.* (Kutologoi east of

AkKorum pass, July 13, 1908—Sap.; "Oigur, Kakkul', Upper Kobd. Daingol lake, Ak-Korum pass, Talnor, Kutologoi, Saksai, Chigirtei, Bzau-Kul', Ulan-Daba"—Sap. l.c.; Tsastu-Bogdo, upper Dzuilin-Gola, 3000–3400 m, June 24; Akhuntyin-Daba along Delyun-Kudzhurtu road, 3050 m, July 2; upper Bulgan river, Ioltyn-Gola valley near Kudzhurtu settlement, Artelin-Sala creek valley 3 km beyond settlement, July 3; upper Buyantu river, Chigirtei-Gol 12 km beyond lake, nor. slope of Chigirtei-Ula, 2800 m, July 4—1971, Grub., Ulzij. et al.).

General distribution: West. Sib. (Altay).

Section 3. **Mesogaea** Bge.

34. O. cachemiriana Camb. in Jacquem. Voy. Ind. 4, Bot. (1844) 38; Bge. Sp. Oxytr. (1874) 43; (err. kashemiriana) excl. pl. tianschan.; Baker in Hook. f. Fl. Brit. Ind. 2 (1876) 139; Fl. Xizang. 2 (1985) 864. —Ic.: Jacquem. l.c., tab. 44; Fl. Xizang. 2, tab. 285, fig. 15–21.

Described from Kashmir. Type in Paris (P).

In small coastal meadows in high mountain areas, 4100–4300 m alt.

IIIB. Tibet: *Chang Tang* ("Getszi"—Fl. Xizang. l.c.), *South.* (Ali: "Chzhada"—Fl. Xizang. l.c.).

General distribution: Himalayas (Kashmir).

Note. Did not see authentic specimens of this species.

35. O. cana Bge. in Bull. Soc. natur. Moscou, 39, 2 (1866) 3; ej. Sp. Oxytr. (1874) 43; Vassilcz. and B. Fedtsch. in Fl. SSSR, 13 (1948) 45; Fl. Kazakhst. 5 (1961) 337; Filim. in Opred. rast. Sr. Azii [Key to Plants of Mid. Asia] 7 (1983) 345; Claves pl. Xinjiang. 3 (1985) 87.

Described from East. Kazakhstan (Jung. Ala Tau). Type in St.-Petersburg (LE).

On rubbly and rocky steppe montane slopes.

IIA. Junggar: *Tarb.* ("Dachen" [Chuguchak]—"Qinhe" [Chingil']—Claves pl. Xinjiang. l.c.), *Jung. Ala Tau* ?

General distribution: Jung.-Tarb.

Note. The report of presence of this species somewhere in Tarbagatai on way to Chingill' from Chuguchak cited in the Chinese "Key to Plants of Sinkiang" (Claves pl. Xinjiangensium) calls for substantiation. Reliable reports of this species are known so far only from Jung. Ala Tau.

36. O. chorgossica Vass. in Not. Syst. (Leningrad) 20 (1960) 232; Vassilcz. and B. Fedtsch. in Fl. SSSR, 13 (1948) 47, descr. ross.; Fl. Kazakhst. 5 (1961) 337; Filim. in Opred. rast. Sr. Azii [Key to Plants of Mid. Asia] 7 (1983) 345; Claves pl. Xinjiang. 3 (1985) 87.

Described from East. Kazakhstan (Jung. Ala Tau: upper Khorgos river). Type in St.-Petersburg (LE).

On rocky steppe slopes of mountains, pebble beds of rivers in middle belt of mountains.

IIA. Junggar: *Jung. Ala Tau* ("Khorgos town"—Claves pl. Xinjiang. l.c.).
General distribution: Jung.-Tarb. (Jung. Ala Tau).

33   **37. O. deflexa** (Pall.) DC. Astrag. (1802) 77; Ledeb. Fl. alt. 3 (1831) 282; Bge. Sp. Oxytr. (1874) 39; Saposhn. Mong. Alt. (1911) 363; Kryl. Fl. Zap. Sib. 7 (1933) 1726; Peter-Stib. in Acta Horti Gotob. 12 (1937) 72; Vassilcz. and B. Fedtsch. in Fl. SSSR, 13 (1948) 43; Boriss. in Fl. Zabaik. [Flora of Transbaikalia] 6 (1954) 602; Grub. Konsp. fl. MNR (1955) 189; Fl. Kazakhst. 5 (1961) 338; Ulzij. in Issl. fl. i rastit. MNR [Study of Flora and Vegetation of Mongolian People's Republic] 1 (1979) 113; id. in Grub. Opred. rast. Mong. [Key to Plants of Mongolia] (1982) 165; Opred. rast. Tuv. ASSR [Key to Plants of Tuva Autonomous Soviet Socialist Republic] (1984) 150; Claves pl. Xinjiang. 3 (1985) 86; Pl. vasc. Helanshan. (1986) 157. —*Astragalus deflexus* Pall. in Acta Ac. Sci. Petrop. 2 (1779) 268. —*A. retroflexus* Pall. Astrag. (1800) 33. —Ic.: Pall. (1779) l.c. tab. 15; Pall. Astrag. tab. 27; Fl. Zabaik. [Flora of Transbaikalia] 6, fig. 306.

Described from East. Siberia (Fore Baikal). Type in London (BM) or Berlin (B). Plate II, fig. 3.

In coastal meadows and pebble beds, sedge-*Cobresia* wastelands and small alpine meadows, larch groves, in forest and upper belt of mountains, 2200–4200 m alt.

IA. Mongolia: *Khobd.* ("Kalgutty"—Sap. l.c.; Turgen' mountain range, Turgen'-Gola gorge 5 km beyond estuary, Aug, 16, 1971—Grub., Ulzij. et al.), *Mong. Alt.* (Datu-Daban pass, July 17, 1874—Klem.; bank of river east of Sumdairyk between moraines, July 30, 1908—Sap.; "Daingol, Talnor"—Saposhn. l.c.; Khara-Adzarga mountain range, nor. slope around Khairkhan-Duru, Aug. 25, 1930—Pob.; Taishiri-Ula mountain range, south. slope in upper Shine-Usu river, 2380–2450 m, June 18, 1971—Grub. Ulzij. et al.; 2 km south of pass along road to Kudzhurtu, upper Ulagchin-Gol, 2840 m, July 2; Tsastu-Bogdo mountain range, upper Dzuilin-Gola, 3000–3400 m, July 24–1971, Grub., Ulzij. et al.), *Gobi-Alt.* (Dzun-Saikhan mountain range along Tsagan-Gol river, Aug. 23, 1931—Ik.-Gal.; Dundu- and Dzun-Saikhan mountain ranges, Khurmein somon, July–Aug. 1930—M. Simukova), *Depr. Lakes* (Dzun-Dzhargalant mountain range, Ulyastyn-Gola gorge, 1850–2800 m, June 28, 1971—Grub.; Ulzij. et al.), *Alash. Gobi* ("Alashan mountain range; Suyuigou gorge and Lin'gun settlement, 2200 m"Pl. vasc. Helanshan. l.c.).

IIIA. Qinghai: *Nanshan* (Humboldt mountain range, nor. slope of Kuku-Usu area, 2750–3050 m, June 3, 1899—Rob.), *Amdo* ("Ba valley, June 1926, No. 14243—Rock"—Peter-Stib. l.c.).

IIIB. Tibet: *Weitzan* (mountains along Talachyu river, 4100 m, June 24 [July 6]; along bank of Razboinich'ei river, 4100 m, July 14 [26], 1884—Przew.).

General distribution: West. Sib. (Altay), East. Sib., Far East (Okhot.), Nor. Mong. (Fore Hubs., Hent., Hang.), China (Altay, North-West), Nor. America.

38. O. gerzeensis P.C. Li in Fl. Xizang. 2 (1985) 859.
Described from Tibet (Chang Tang). Type in Beijing (PE).
In alpine meadows.

IIIB. Tibet: *Chang Tang* ("Gerze, alt. 5200 m, on meadows of hills, Sept. 14, 1976, No. 12567—Qinghai-Xizang Compl. Exp., typus!").
General distribution: endemic.

Note. This is perhaps a variety of *O. melanocalyx* Bge. differing in dense white-silky-villous pubescence of leaflets.

39. **O. glabra** (Lam.) DC. Astrag. (1802) 95; Bge. Sp. Oxytr. (1874) 40; Danguy in Bull. Mus. nat. hist. natur. 17, 4 (1911) 270; Saposhn. Mong. Alt. (1911) 363; Hedin, S. Tibet, 6, 3 (1922) 62; Kryl. Fl. Zap. Sib. 7 (1933) 1727; Peter-Stib. in Acta Horti Gotob. 12 (1937) 72; Persson in Bot. Notiser, 4 (1938) 291; Hao in Bot. Jahrb. 68 (1938) 614; Walker in Contrib. U.S. Nat. Herb. 28 (1941) 638; Vassilcz. and B. Fedtsch. in Fl. SSSR, 13 (1948) 41; Boriss. in Fl. Zabaik. [Flora of Transbaikalia] 6 (1954) 600; Grub. Konsp. fl. MNR (1955) 189; Fl. Kirgiz. 7 (1957) 373; Fl. Kazakhst. 5 (1961) 333; Fl. Intramong. 3 (1977) 218; Fl. Tadzh. 5 (1978) 451; Ulzij. in Issl. fl. i rast. MNR [Study of Flora and Vegetation of Mongolian People's Republic] 1 (1979) 113; id. in Grub. Opred. rast. Mong. [Key to Plants of Mongolia] (1982) 165; Filim. in Opred. rast. Sr. Azii [Key to Plants of Mid. Asia] 7 (1983) 344; Opred. rast. Tuv. ASSR [Key to Plants of Tuva Autonomous Soviet Socialist Republic] (1984) 150; Claves pl. Xinjiang. 3 (1985) 88; Fl. Xizang. 2 (1985) 859. —*O. diffusa* Ledeb. Fl. Alt. 3 (1831) 281; ej. Fl. Ross. 1 (1843) 585; Henderson and Hume, Lahore to Jarkand (1873) 318; Baker in Hook. f. Fl. Brit. India, 2 (1876) 140. —*O. drakeana* Franch. Pl. David. 1
34 (1884) 89. —*O. puberula* Boriss. in Tr. Tadzh. bazy AN SSSR, 2 (1936) 169; Vassilcz. and B. Fedtsch. in Fl. SSSR, 13 (1948) 43; Fl. Kirgiz. 7 (1957) 373; Fl. Kazakhst. 5 (1961) 334; Fl. Tadzh. 5 (1978) 453; Filim. in Opred. rast. Sr. Azii [Key to Plants of Mid. Asia] 7 (1983) 344; Claves pl. Xinjiang. 3 (1985) 85. —*O. glareosa* Vass. in Not. Syst. (Leningrad) 20 (1960) 232; Vassilcz. and B. Fedtsch. in Fl. SSSR, 13 (1948) 49 descr. ross.; Fl. Kazakhst. 5 (1961) 336; Ulzij. in Issl. fl. i rast. MNR [Study of Flora and Plants of Mongolian People's Republic] 1 (1979) 114; id. in Grub. Opred. rast. Mong. [Key to Plants of Mongolia] (1982) 165; Filim. in Opred. rast. Sr. Azii [Key to Plants of Mid. Asia] 7 (1983) 346; Claves pl. Xinjiang. 3 (1985) 88. —*O. salina* Vass. in Not. Syst. (Leningrad) 20 (1960) 234; Vassilcz. and B. Fedtsch. in Fl. SSSR, 13 (1948) 50, descr. ross.; Boriss. in Fl. Zabaik. [Flora of Transbaikalia] 6 (1954) 602; Ulzij. in Issl. fl. i rast. MNR [Study of Flora and Vegetation of Mongolian People's Republic] 1 (1979) 115; id. in Grub. Opred. rast. Mong. [Key to Plants of Mongolia] (1982) 165. —*Astragalus glaber* Lam. Encycl. meth. bot. 1 (1783) 525. —**Ic.:** DC. l.c. tab. 8; Ledeb. Ic. pl. fl. ross. 5, tab. 451 (sub nom. *O. diffusa*); Franch. l.c. tab. 12; Fl. Kazakhst. 5, Plate 43, fig. 1; Grub. Opred. rast. Mong. [Key to Plants of Mongolia] Plate 88, fig. 403 (sub nom. *O. salina*); Fl. Intramong. 3, tab. 110, fig. 1–10.

Described from Siberia (?). Type in Paris (P).

On solonchak, solonetzic and swampy meadows along banks of rivers, brooks and lakes and valley floors, chee grass thickets, iris groves, on pebble beds and sandy shoals of rivers, small meadows around springs and banks of irrigation channels.

IA. Mongolia: all regions.
IB. Kashgar: *Nor., West., South., East.*
IC. Qaidam: *plains* and *mountains.*
IIA. Junggar: all regions.
IIIA. Qinghai: *Nanshan, Amdo.*
IIIB. Tibet: all regions.
IIIC. Pamir.
General distribution: Aralo-Casp., Fore Balkh., Jung.-Tarb., Nor. and Cen. Tien Shan, East. Pam.; Mid. Asia, West, Sib., East. Sib., Nor. Mong., China (North-West).

Note. Unusually polymorphic species, showing considerable variation under different habitat conditions in the overall plant size, extent of development and growth form of stem (erect to procumbent), size of leaflets, flowers and fruits, extent of pubescence of plant as a whole (from glabrous to pubescent) including pods. This was cited by Ledebour himself who distinguished α *elongata* and β *pumila* in his *O. diffusa* as well as by A. Bunge.

Plants with patently-pilose pods (*O. puberula*) are found sporadically all over the distribution range of the species, from Tajikistan to Mongolia.

There is no real distinction also between creeping small forms—*O. glareosa* and *O. salina*—since the length of keel cusp, the basis of their differentiation, is highly variable while these forms themselves are morphologically not differentiated from the main form and have no distinct distribution range.

40. O. gorbunovii Boriss. in Tr. Tadzh. bazy AN SSSR, 2 (1936) 168; Vassilcz. and B. Fedtsch. in Fl. SSSR, 13 (1948) 49; Fl. Kazakhst. 5 (1961) 336; Ikonn. Opred. rast. Pamira [Key to Plants of Pamir] (1963) 174; Filim. in Opred. rast. Sr. Azii [Key to Plants of Mid. Asia] 7 (1983) 345; Claves pl. Xinjiang. 3 (1985) 87. —*O. goloskokovii* Bait. in Fl. Kazakhst. 5 (1961) 493, 334.

Described from Mid. Asia (Badakhshan). Type in St.-Petersburg (LE).

In coastal meadows and pebble beds, shoals, banks of brooks and irrigation channels, from steppe to upper belt of mountains.

IB. Kashgar: *Nor.* (Uchturfan, Karagailik gorge, June 17 and 18, 1908—Divn.), *West.* (King Tau mountain range, nor. slope near Kosh-Kulak settlement 40–50 km south-west of Upal oasis, June 10, 1959—Yun. and I-f. Yuan'), *South.* (nor. slope of Russky mountain range near Achan village, 2300 m, June 10 [22]; Keriya mountain range, Nura river gorge, 3150 m, July 11 [23]; foothills of Keriya mountain range along Karatash river, 2450 m, July 26 [Aug. 7]—1885, Przew.; Polu, May 25, 1890—Grombch.; nor. slope of Russky mountain range near Karasai village, 3050 m, July 8; upper Cherchen river, Yan'dash area, Aug. 7, 1890—Rob.).

35   **IIIC. pamir** (Charlym river, along bank, June 21, 1909—Divn.; Issyk-Su river, 3000–3100 m, June 21, 1942—Serp.).

General distribution: Jung. Ala Tau, Nor. and Cen. Tien Shan, East. Pam.; Mid. Asia (Pam.-Alay).

41. O. gueldenstaedtioides Ulbr. in Bot. Jahrb. 36, Beibl. 82 (1905) 65; Peter-Stib. in Acta Horti Gotob. 12 (1937) 73.

Described from North-West China (Shenxi). Type in Berlin (B) ?

On Steppe and meadow slopes of mountains, solonetzic coastal meadows and valley floors.

IIIA. **Qinghai:** *Nanshan* (along Kuku-Usu river, alpine belt, 2450–2600 m, June 28 [July 10]; same site, July 4 [16]; [Machan-Ula] alpine belt, 3350–3650 m, July 11 [23] 1879; on nor. foothill of South. Kukunor mountain range, 3050 m, June 23 [July 5] 1880—Przew.; Sharagol'dzhin river, Sungi-Nor area, about 3050 m, July 9, 1894—Rob.; 70 km south-east of Chzhan'e town, Matisy temple, 2600 m, July 12, 1958; 24 km south of Xining, 2650 m, Aug. 4, 1959—Petr.), **Amdo** (Khagomi on upper Huang He, 2150 m, June 23 [July 5] 1880—Przew.).

IIIB. **Tibet:** *Weitzan* (Burkhan-Budda mountain range, nor. slope, Khatu river gorge, 3200 m, June 15, 1901—Lad.).

General distribution: China (North-West).

42. O. hirsutiuscula Freyn in Bull. Herb. Boiss. 2 ser. 5 (1905) 1021; Vassilcz. and B. Fedtsch. in Fl. SSSR, 13 (1948) 48; Fl. Kirgiz. 7 (1957) 375; Ikonnik. Opred. rast. Pamira [Key to Plants of Pamir] (1963) 172; Filim. in Opred. rast. Sr. Azii [Key to Plants of Mid. Asia] 7 (1983) 345; Claves pl. Xinjiang. 3 (1985) 87.—*O. glabra* var. *pamirica* B. Fedtsch. in Acta Horti Petrop. 21, 3 (1903) 130.

Described from Pamir (Yashil'-Kul' lake). Type in St.-Petersburg (LE).

On banks of rivers and rivulets, coastal meadows, steppe to alpine belt of mountains.

IIA. **Junggar:** *Tien Shan* (B. Yuldus, 2450 m, steppe, Aug. 3, 1893—Rob.).

IIIA. **Qinghai:** *Nanshan* (Humboldt mountain range, Shargol'dzhin river valley, 3350 m, June 15; same site, in Baga-Bulak area, 3200 m, June 15; Yamatyn-Umru hills, 3650 m, July 21—1894, Rob.).

IIIC. **Pamir** (Gumbus river, Chon-Terek area, June 7, 1909—Divn.; "Tashkurgan"- Claves pl. Xinjiang. l.c.).

General distribution: East. Pam., Mid. Asia (Pam.-Alay).

43. O. kansuensis Bge. in Mem. Ac. Sci. St.-Petersb. 22, 1 (1874) 38; Forbes and Hemsl. Index Fl. Sin. 3 (1905) 497; Danguy in Bull. Mus. nat. hist. natur. 17 (1911) 270; Peter-Stib. in Acta Horti Gotob. 12 (1937) 74; Hao in Bot. Jahrb. 68 (1938) 613; Fl. Xizang. 2 (1985) 858. —*O. leucocephala* Ulbr. in Notizbl. Bot. Gart. Berlin, 3 (1902) 193.

Described from Qinghai (Nanshan). Type in St.-Petersburg (LE). Plate I, fig. 1. Map 1.

In montane steppes, spruce forests, among shrubs, on banks of rivers in upper mountain belt.

IC. **Qaidam:** *mountains* (Dulan-Khit temple, 3350 m, in spruce forest, Aug. 9, 1901—Lad.).

IIIA. Qinghai: *Nanshan* (Jugum alpinum secus fl. Tetung, regio alpina, July 10–24, 1872, No. 275—Przew., typus!; [upper Loto river in North-Tetung mountain range], Cherik pass, Aug. 8; nor. bank of Kuku-Nor, Dege-chyu river, Aug. 9—1890, Gr.-Grzh.; nor. slope of Humboldt mountain range, June 9, 1894—Rob.; Tsilyan'shan', Matisy temple, 2600 m, nor. slope of mountain range, July 12; east. extremity of Nanshan 25 km south of Gulan' town, rolling mountain slopes, 2435 m, Aug. 12; 37 km south of Gulan' town, crossing along road to Lyan'chzhou, 2950 m, Aug. 12—1958, Petr.; Mon'yuan', Datunkhe river near stud farm, 2800 m, nor.-east. slope, 1958—Dolgushin; pass 86 km west of Xining, Aug. 5; 108 km west of Xining and 6 km west of Daudankhe settlement, 3400 m, Aug. 5, 1959—Petr.; "Sining-Fou, alt. 2400 m, July 13, 1908, Vaillant"—Danguy l.c.; "Sining-fu bei Schangwu-chuang, 2800–3000 m, 1930 "—Hao l.c.), *Amdo* (Dzhakhar-Dzhargyn mountains, 3200–3500 m, June 9 [21] 1880—Przew.; "Ba Valley, No. 14249; Jupar range, No. 14343, June 1926, Rock"—Peter-Stib. l.c.).

IIIB. Tibet: *Chang Tang* ("Mandarlik, 3437 m, July 1900", Hedin—Ulbr. l.c. 1922), *Weitzan* (Burkhan-Budda mountain range, nor. slope, Khatu gorge, 3350–3650 m, June 12, 1901—Lad.). *South.* ("Lhasa"—Fl. Xizang. l.c. ?).

General distribution: China (North-West: Kansu; South-West: Sichuan).

36

Note. 1. The species is very close to *O. meinshausenii* Schrenk. The report in Fl. Xizang. l.c. to the occurrence of this species in South. Tibet ("Lhasa") arouses doubt.

2. The separation of yellow-coloured crazyweeds with developed stems into separate section *Chrysantha* Vass. of subgenus *Eumorpha*, in our opinion, is not justified. Primarily, they belong more readily to subgenus *Oxytropis* since their pods bear false lower wall formed by the depression of ventral suture and, essentially, are unilocular. A. Bunge himself drew attention to this feature (Sp. Oxytr.: 57) and placed some of these species (*O. kansuensis, O. meinshausenii*) straightway in section *Mesogaea* and some in section *Ortholoma* (*O. ochrocephala, O. pilosa*). I.T. Vassilczenko also reported this feature (Fl. SSSR, 13; 104). Moreover, yellow-coloured species essentially differ in no respect other than the colour of corolla from blue-coloured species of the same section *Mesogaea*. Thus, *O. kansuensis* has its own Chinese blue-coloured analogue *O. gueldenstaedtioides* Ulbr. with fruits of same form. Apart from those included by I.T. Vassilczenko (Fl. SSSR, 13: 103) in section *Chrysantha*, the group of such yellow-coloured crazyweeds is represented in China by *O. ochrocephala* Bge. as well, apart from the above-cited *O. kansuensis* Bge.

44. O. lapponica (Wahl.) J. Gay in Flora, 10, 2 (1827) 30; Gaud. Fl. Helv. 4 (1829) 543; Bge. Sp. Oxytr. (1874) 8; Kryl. Fl. Zap. Sib. 7 (1933) 1721; Vassilcz. and B. Fedtsch. in Fl. SSSR, 13 (1948) 17; Grub. Konsp. fl. MNR (1955) 190; Fl. Kirgiz. 7 (1957) 365; Fl. Kazakhst. 5 (1961) 339; Fl. Tadzh. 5 (1978) 450; Ulzij. in Issled. fl. i rastit. MNR [Study of Flora and Vegetation of Mongolian People's Republic] 1 (1979) 112; id. in Grub. Opred. rast. Mong. [Key to Plants of Mongolia] (1982) 165; Filim. in Opred. rast. Sr. Azii; [Key to Plants of Mid. Asia] 7 (1983) 346; Opred. rast. Tuv. ASSR [Key to Plants of Tuva Autonomous Soviet Socialist Republic] (1984) 150; Claves pl. Xinjiang. 3 (1985) 89; Fl. Xizang. 2 (1985) 850. —*Phaca lapponica* Wahl. Veg. Helvet. (1813) 131 in nota. —**Ic.**: J. Gay l.c. Suppl. tab. 20; Reichb. Ic. Fl. Germ. 22, tab. 173; Fl. Kazakhst. 5, Plate 48, fig. 5; Claves pl. Xinjiang. 3, tab. 6, fig. 4.

Described from Switzerland. Type in London (K).

On coastal and wet meadows, banks of rivers and rivulets, moraines and talus, meadow slopes, larch forests in middle and upper belts of mountains.

IA. Mongolia: *Khobd.* (Kharkira river valley beyond Turguni estuary, July 21, 1879; Katu river, right tributary of Bukon'-Bere river, in lower part of gorge, June 15, 1879—Pot.; Turgen' mountain range, Turgen'-Gola gorge 5 km from estuary, July 16, 1971—Grub., Ulzij. et al.; Elyn-Ama area 21 km nor.-west of Ulan-Daba pass, nor. slope, July 16, 1977—Karam., Sanczir et al), *Mong. Alt.* (Taishiri-Ula, July 13 and 15, 1877—Pot.; bank of Naryn brook around pass, July 21, 1896—Klem.; Taishiri-Ula, nor. slope, July 12, 1945—Yun.; upper Bulgan river 3 km beyond Kudzhurtu settlement, Artelin-Sala creek valley, July 3, 1971—Grub., Ulzij. et al.; Ulyastei-Gol valley, Shazdgat-Nuru mountain range, 2100 m, June 27, 1973—Golubkova and Tsogt).

IB. Kashgar: *West.* (Sarykol mountain range, Bostan-Terek area, alp. belt, Aug. 11, 1929—Pop.).

IIA. Junggar: *Tien Shan* (Sairam lake, July 20; Dzhagastai pass, 3050 m, Aug. 11—1877, A. Reg.; valley of midcourse of Muzart-Khan-Dzhailyau, 2150–2450 m, Aug. 17; Muzart gorge, 3050–3500 m, Aug. 17, 1877; Burkhantau [Tekesa basin], June 6; valley north of Sairam lake, July 12; west of Sairam lake, July 20; Sairam-Kyzimchek, July 23; Arystan on Ili river, Aug. 7—1878, Fet.; between Sumbe pass and Kassan, 2150–2450 m, June 22; Kok-Tyube, 2150–2750 m, June 23; Chapchal gorge in Dzhagastai mountains, 2450 m, June 28; Bogdo hill nor. of Kul'dzha, 3050 m, July 25; Kok-Kamyr mountains, 2150–2450 m, Aug. 31—1878, A. Reg.; Kumbel', 2750–3050 m, May 31; Chunkur-Daban pass nor. of Borborogusun, 3050 m, June 13; along Borborogusun river, 2750 m, June 15; Yulty-Aryslan, 2150–2450 m, July 7; Aryslan, 2750–3050 m, July 12 and 16; Mengute, 3050–3350 m, Aug. 9; Kash river source, 3050–3350 m, Aug. 12; Kash river, 2750 m, Aug. 17–18; Kunges river, 2150–2450 m, Aug. 27—1879, A. Reg.; oberstes Dschanart Tal, June 14–17, 1903; Lager am sudrande des Bogdo-Ola, Aug. 26–29, 1908—Merzb.).

IIIB. Tibet: *Chang Tang* (Keriya mountain range, Kyuk-Egil' gorge, 3800–3950 m, June 28 [July 10]; same site, along Khan-Yut river, 3650 m, July 13 [25]—1885, Przew.; Russky mountain range, south. slope, Aksu river, 3650 m, July 2; same site, nor. slope, Karasai village, 3050 m, July 8—1890, Rob.).

General distribution: Jung.-Tarb., Nor. and Cen. Tien Shan, East. Pam.; Arct. (Europ.), Europe (mountains), Caucasus (Glavnyi mountain range), Mid. Asia (West. Tien Shan, Pam.-Alay), West. Sib. (Altay), Nor. Mong. (Hang.), China (Altay).

Note. The reference of Diels [in Filchner, Wiss. Ergebn. 10 (1908) 254] to the report of this species in the vicinity of Xining town (Qinghai) perhaps pertains to its closely related species *O. melanocalyx* Bge.

45. O. meinshausenii Schrenk in Bull. Ac. Sci. St.-Petersb, 10 (1842) 254; id. in Fisch. et Mey. Enum. pl. nov. 2 (1842) 49; Ledeb. Fl. Ross. 1 (1843) 786; Bge. Sp. Oxytr. (1874) 45; Vassilcz. and B. Fedtsch. in Fl. SSSR, 13 (1948) 103; Fl. Kirgiz. 7 (1957) 377; Fl. Kazakhst. 5 (1961) 388; Filim. in Opred. rast. Sr. Azii [Key to Plants of Mid. Asia] 7 (1983) 350; Claves pl. Xinjiang. 3 (1985) 105. —**Ic.**: Fl. Kazakhst. 5, Plate 43, fig. 3.

Described from East. Kazakhstan (Jung. Ala Tau). Type in St.-Petersburg (LE). Plate III, fig. 1.

Along forest borders and meadows, coastal meadows and meadow slopes in midbelt of mountains.

IIA. Junggar: *Tien Shan* (west. tributary of Aryslyn, 2450–2750 m, July 8; estuary of Aryslyn gorge, 2450 m, July 10; Mengute, 2750 m, Aug. 3; Algoi, 1850–2450 m, Sept. 12—1879, A. Reg.; prope Musart superior, 1886—Krassnow; Kelisu, 1700 m, No. 1877, July 17; between Kelisu and Daban, 2200 m, No. 1944, July 18—1957, Kuan; south of Nyutsyuan'tsz, No. 632, July 18, 1957—Shen: Datszymyao village, No. 1714, July 2, 1957—Kuan: 8–10 km nor. of Chzhaosu, 2110 m, Aug. 15; between Chzhaosu and Tekes, No. 3585, Aug. 16—1957, Kuan; Narat mountains on Kunges, No. 6592, Aug. 7, 1958—Lee and Chu; nor. foothills of Narat mountain range, descending into Tsanma valley, Aug. 7, 1958—Yun. and I.-f. Yuan'; "Sin'yuan', Gunlyu"—Fl. Xizang. l.c.).

General distribution: Jung.-Tarb. (Jung. Ala tau), Nor and Cen. Tien Shan.

46. **O. melanocalyx** Bge. in Mem. Ac. Sci. St.-Petersb. ser. 7, 22, I (1874) 8; Forbes and Hemsley, Index Fl. Sin. 3 (1905) 497; Peter-Stib. in Acta Horti Gotob. 12 (1937) 70; Hao in Bot. Jahrb. 68 (1938) 613; Walker in Contribs. U.S. Nat. Herb. 28 (1941) 638; Fl. Xizang. 2 (1985) 850. —*O. lapponica* auct. non Gay: Ulbr. in Bot. Jahrb. 36, Beibl. 82 (1905) 65. —*O. montana* auct. non DC.: Ulbr. ibid and in Feddes Repert., Beih. 12 (1922) 424. —Ic.: Fl. Xizang. 2, tab. 279, fig. 1–7.

Described from China (Qinghai). Type in St.-Petersburg (LE).

Along river banks, on pebble beds and meadows, shoals, alpine meadows, up to 4500 m alt.

IIIA. Qinghai: *Nanshan* (In regione silvestri jugi ad meridiem fl. Tetung [ad monaster. Tcheibsen], July 4–9 [16–21] 1872—Przew. typus!; "Hsin-Cheng, west of Lanchow, June 1923, R.C. Ching"—Walker l.c.), *Amdo* (in pratis alpinis montium Mudshik, 3200 m, June 12 [24]; summa regione alpine montium Mudshik, 3950 m, June 14 [26] 1880—Przew.; "Amne-Matchin, um 4500 m"—Hao l.c.).

IIIB. Tibet: *Weitzan* (Jugo inter fl. Hoangho et Jangtze, May 30 [June 11]; regio alpina superior secus fl. Jangtze [Dytshiu] June 8 [20]; ad affluentem dextrum fl. Bytshiu, 4200 m, July 1 [13]; montes ad fl. Talatschiu, 4000 m, July 14 [26]—1884, Przew.; Dzhagyn-Gol river, 4200 m, No. 169a, 169b, July 1, 1900—Lad.).

General distribution: China (North-West: Shenxi, Kansu; South-West).

Note. References to the report of this species in Chang Tang [Diels in Filchner, Wiss. Ergebn. 10 (1908) 254; Ostenfeld and Paulsen in Hedin, S. Tibet, 6, 3 (1922) 63] arouse doubt: these citations probably refer to *O. lapponica*.

47. **O. ochrocephala** Bge. in Mem. Ac. Sci. St.-Petersb. 7 ser. 22, I (1874) 57; Forbes and Hemsl. Index Fl. Sin. 3 (1905) 497; Peter-Stib. in Acta Horti Gotob. 12 (1937) 75; Hao in Bot. Jahrb. 68 (1938) 614; Fl. Xizang. 2 (1985) 858. —*O lapponica* var. *xanthantha* Baker in Hook. Fl. Brit. Ind. 2 (1876) 137.

38   Described from Qinghai (Nanshan). Type in St.-Petersburg (LE).

On steppe and meadow slopes, scrubs. coastal meadows and pebble beds, small alpine meadows in middle and upper belts of mountains.

IA. Mongolia: *Khesi* (Loukhushan' mountain range, July 17, 1908—Czet.).

IIIA. Qinghai: *Nanshan* (prov. Kansu, pars orientalis, med. June 1872, No. 109—Przew., typus!; non procul a Lantscheu July 14 [26] 1875—Piassezki; ad fl. Yussun-Chatyma, 2750–3050 m, July 12 [24]; regio collina versus monasterium Tschoibsen,

July 13 [25] 1880—Przew.; valley of Tyan'chyg river near Khulanchir village [near Xining town], April 16, 1885—Pot.; nor. slope of Humboldt mountain range around Blagodatnyi spring, 3050 m, June 9, 1894; same site, Chonsai gorge near pass, about 3650 m, July 22, 1895—Rob.; Fynfyilin' pass, south-west. slope, July 20, 1909—Czet.; pass through east. fringe of Nanshan 75 km east of Yunchan town along road to Lanzhou, Oct. 8, 1957—Yun. ?; 25 km south of Gulan town, low mountains, 2435 m, Aug. 12; Tsilyan'-Shan' mountains, Matisy temple, 2000 m, in mountain valley, July 12—1958; 24 km south of Xining, 2650 m, slopes of hillocks, Aug. 4, 1959—Petr.; "auf dem ostlichen Nanschan, 2800–3200 m; auf der Ebene Schachuyi, um 3800 m"-Hao l.c.), *Amdo* (ad fl. Mudshikche, S. a Huidui, 2750–2900 m, June 4–16; prope alpem Dshachar-Dsargyn, 3200–3500 m, June 9 [21]—1880, Przew.; "J. Rock No. 12925, 13146"—Peter-Stib. l.c. ?).

IIIB. Tibet: *Chang Tang* (Keria-Jugum, 3200 m, Tschi-wei, July 3 [15]; ibid, Uluk-Atschik, 3350–3650 m, July 24 [Aug. 5]—1885, Przew.) *Weitzan* (fauce ad rivulum prope cacumen jugi inter. fl. Hoangho et Jangtze, May 30 [June 11]; montes ad fl. Bytschu, June 5 [17]; ad fl. Ladronum, 4100 m, July 14 [26]—1884, Przew.; Burkhan-Budda mountain range, nor. slope, Nomokhun gorge, 4400 m, May 24; stream connecting Russkoe lake with Ekspeditsii lake, 4100 m, June 29—1900; valley of Makhmukhchyu brook, 4100 m, May 21; Burkhan-Budda mountain range, south. slope toward Alyk-Nor lake, 2800 m, June 9—1901, Lad.), *South* ("Lhasa and south. Tibet, 3000–5000 m"—Fl. Xizang. l.c.).

General distribution: China (North-West: Kansu; South-West: Sychuang, Kam), Himalayas (Kashmir).

Note. Very closely related to and, usually difficult to distinguish from *O. kansuensis* Bge. The Herbarium of the Komarov Botanical Institute of the Russian Academy of Sciences has a specimen which no doubt belongs to this species with a hand-written label "In locis subarenosis Mongolia chinensis, 1831" with no record of data of collection or the name of collector which could not be deciphered from the hand-writing. This is evidently an error that occurred during the mounting of the plant caused by the switching of labels.

48. **O. ochroleuca** Bge. in Bull. Soc natur. Moscou, 39, 2 (1866) 6; Bge. Sp. Oxytr. (1874) 10; Vassilcz. and B. Fedtsch. in Fl. SSSR, 13 (1948) 44; Fl. Kirgiz. 7 (1957) 374; Fl. Kazakhst. 5 (1961) 339; Filim. in Opred rast. Sr. Azii [Key to Plants of Mid. Asia] 7 (1983) 347; Claves pl. Xinjiang. 3 (1985) 86. —Ic.: Gartenfl. 33, 2, tab. 1154.

Described from Cen. Tien Shan (Karkara river). Type in St.-Petersburg (LE).

In meadows and meadow slopes, spruce forests and their borders, coastal meadows and pebble beds, rocky slopes in middle and upper belts of mountains.

IIA. Junggar: *Tarb.* ("Dachen-Qinhe"-Claves pl. Xinjiang. l.c.), *Tien Shan* (Talki river gorge, July 18; south-east. coast of Sairam lake, July 22; Dzhagastai, Aug. 8; valley of Muzart river below pass, 2750–3200 m, Aug. 19—1877, A. Reg.; Sairam lake, 2600–2750 m, July 12, 1878—Fet.; Kassan river, 2150 m, June 22; Kok-Kamyr mountains, 1850–2150 m, July 27—1878; Mengute, 2150 m, July 27, 1879—A. Reg.; B. Yuldus, 2750 m, Aug. 3, 1893—Rob.; nordliches Kuikonik Tal nahe der Mundung in das Juldus Tal, June 17–19, 1908—Merzb.; Manas river basin, Ulan-Usu river valley 1–2 km before confluence of right tributary Koisu, July 17, 1957—Yun. and I-f. Yuan'; around Danyu, nor. slopes, No. 1454, July 17, 1957—Kuan; south. slope of Narat mountain range descending into B. Yuldus basin from Dachit pass to Bain-Bulak,

Aug. 8, 1958—Yun. and I-f. Yuan'; Ulastai, Mal. Yuldus, 2500 m, No. 6316, Aug. 2, 1958—Lee and Chu.

General distribution: Jung.-Tarb., Nor. Tien Shan.

Note. In all probability, this is a hybrid of *O. lapponica* (Wahlb.) Gay × *O. meishausenii* Schrenk since all its characteristics are unstable and numerous intermediates tending toward the latter species are seen; sometimes, stems are numerous and very short as in *O. lapponica* or sometimes long and solitary as in *O. meinshausenii*, and pods erect while the violet colouration of the tip of keel is not visible and flowers are indistinguishable from those of *O. meinshausenii*, and so on.

39   49. **O. pilosa** (L.) DC. Astrag. (1802) 73; ej. Prodr. 2 (1825) 280; Ledeb. Fl. Ross. 1 (1843) 584; Bge. Sp. Oxytr. (1874) 58; Kryl. Fl. Zap. Sib. 7 (1933) 1731; Vassilcz. and B. Fedtsch. in Fl. SSSR, 13 (1948) 104; Boriss. in Fl. Zabaik. [Flora of Transbaikalia] 6 (1954) 616; Fl. Kazakhst. 5 (1961) 388; Fl. Tsentr. Sib. [Flora of Central Siberia] 2 (1979) 620; Filim. in Opred. rast. Sr. Azii [Key to Plants of Mid. Asia] 7 (1983) 350; Claves pl. Xinjiang. 3 (1985) 105. —*Astragalus pilosus* L. Sp. pl. (1753) 756; Pall. Sp. Astrag. (1800) 106. —**Ic.:** Pall. l.c. tab. 80; Ledeb. Ic. pl. fl. ross. 5, tab. 451; Fl. Kazakhst. 5, Plate 43, fig. 2.

Described from Europe and Siberia. Type in London (Linn.).

In steppes on plains, trails and foothills, coastal sand.

IIA. Junggar: *Cis-Alt.* ("Altay"—Claves pl. Xinjiang. l.c.), **Dzhark.** "Ili" [Kul'dzha]—Claves pl. Xinjiang. l.c.), **Balkh-Alak.** ("Dachen" [Chuguchak]—Claves pl. Xinjiang, l.c.).

General distribution: Jung.-Tarb.; Europe, Balk.-Asia Minor, Caucasus, West. Sib., East. Sib. (Ang.-Sayan.).

50. **O. platonychia** Bge. in Mem. Ac. Sci. St.-Petersb. 7 ser. 22, I (1874) 44; Vassilcz. and B. Fedtsch. in Fl. SSSR, 13 (1948) 50; Fl. Kirgiz. 7 (1957) 376; Ikonnik. Opred. rast. Pamira [Key to Plants of Pamir] (1963) 171; Fl. Tadzh. 5 (1978) 456; Filim. in Opred. rast. Sr. Azii [Key to Plants of Mid. Asia] 7 (1983) 347; Claves pl. Xinjiang. 3 (1985) 88. —**Ic.:** Fl. SSSR, 13, Plate 8, fig. 3; Ikonnik. l.c. Plate 20, fig. 2, 3.

Described from Mid. Asia (Alay). Type in St.-Petersburg (LE).

On rocky and rubbly slopes, talus, pebble beds of rivers and small meadows in high mountains.

IIIC. Pamir ("Tashkurgan"—Claves pl. Xinjiang. l.c.).
General distribution: East. Pam.; Mid. Asia (Pam.-Alay).

Note. 1. The references to the occurrence of this species in Tien Shan (Fl. Kirgiz., Fl. Kazakhst., Claves pl. Xinjiang.) are not confirmed by reliable herbarium specimens.
2. The placement of this species into separate section *Dolichanthos* Gontsch. is not enough justified.

51. **O. ulzijchutagii** Sancz. in Tr. Inst. bot. AN MNR, 7 (1985) 93; id. in BNMAU Shinzhlekh Ukhaany Adademiin Medee [Proceeding of the Mongolian People's Republic Academy of Sciences], 3 (1985) 52. —**Ic.:** Tr. Inst. bot. AN MNR, 7 : 94, fig. 2.

Described from Mongolia (Mong. Altay). Type in Ulan-Bator (UB).

In wet and marshy low-grass meadows, *Cobresia* thickets, moist rocks in alpine belt.

IA. **Mongolia:** *Khobd.* (Khobdo somon, west. slope of Khoit-Khetsu-Ulan-ula, swampy meadow, July 30, 1977—Karamysheva, Sanczir), *Mong. Alt.* (Tsogt somon, Tsakhar-Khalgany-Nuru mountains, 3200 m, in *Cobresia* thickets on rocks, Aug. 13, 1973, No. 6079—E. Isachenko and E. Rachovskaya, typus!).

General distribution: endemic.

Subgenus II. EUMORPHA (Bge.) Abduss.

Section 1. Eumorpha Bge.

52. O. barkultagi Grub. et Vass. in Novit. Syst. pl. vasc. 24 (1987) 138.

Described from Sinkiang (Tien Shan). Type in St.-Petersburg (LE).

On meadow slopes in forest belt of mountains.

IIA. **Junggar:** *Tien Shan* (Barkul'tag mountain range, Koshety-Daban pass, occasionally on both slopes, on southern slope in gorge along flanks, May 24 [June 5] 1879, No. 131—Przew., typus; Koshety-Daban, nor. slope, forest zone, June 11, 1877—Pot.).

General distribution: endemic.

40   **53. O. cuspidata** Bge. in Mem. Ac. Sci. St.-Petersb. 7, ser. 22, I (1874) 70; Vassilcz. and B. Fedtsch. in Fl. SSSR, 13 (1948) 126; Fl. Kazakhst. 5 (1961) 377; Filim. in Opred. rast. Sr. Azii [Key to Plants of Mid. Asia] 7 (1983) 356; Claves pl. Xinjiang. 3 (1985) 110.

Described from East. Kazakhstan (Jung. Ala Tau). Type in Paris (P).

On rubbly and rocky slopes in middle (forest) belt of mountains.

IIA. **Junggar:** *Cis-Alt.* ("Qinhe" [Chingil']—Claves pl. Xinjiang. l.c.), *Tarb.* ("Dachen"—Claves l.c.).

General distribution: Jung. Ala Tau.

54. O. dschagastaica Grub. et Vass. in Novit. Syst. pl. vasc. 24 (1987) 137.

Described from Sinkiang (Tien Shan). Type in St.-Petersburg (LE).

On meadow slopes in midbelt of mountains.

IIA. **Junggar:** *Tien Shan* (Dzhagastai, June 14, 1884, No. 165—A. Reg., typus!).

General distribution: endemic ?

Note. The data on the label written by K. Winkler is clearly incorrect. In 1884, A. Regel was in Bukhara and Turkmenia and in Mid-June 1878 was on Dzhagastai river in Tien Shan which is south of Kul'dzha.

55. O. macrocarpa Kar. et Kir. in Bull. Soc. natur. Moscou, 15 (1842) 326; Bge. Sp. Oxytr. (1874) 71; Vassilcz. and B. Fedtsch. in Fl. SSSR, 13 (1948) 108; Fl. Kirgiz. 7 (1957) 378; Fl. Kazakhst. 5 (1961) 367 excl. syn.;

Filim. in Opred. rast. Sr. Azii [Key to Plants of Mid. Asia] 7 (1983) 350; Claves pl. Xinjiang. 3 (1985) 109. —*O. robusta* M. Pop. ex Vass. et B. Fedtsch. in Fl. URSS, 13 (1948) 545 and 109; Fl. Kazakhst. 5 (1961) 367.

Described from East. Kazakhstan (Jung. Ala Tau). Type in St.-Petersburg (LE). Plate III, fig. 2.

On rocky slopes and rocks in middle (forest) belt of mountains.

IIA. **Junggar:** *Jung. Ala Tau* ("Dachen -Ili"—Claves pl. Xinjiang. l.c.).

General distribution: Jung.-Tarb., Nor. and Cen. Tien Shan; Mid. Asia (Pam.-Alay, West. Tien Shan).

56. **O. semenovii** Bge. in Bull. Soc. natur. Moscou, 39, 2 (1866) 13; ej. Sp. Oxytr. (1874) 66; Vassilcz. and B. Fedtsch. in Fl. SSSR, 13 (1948) 122; Fl. Kazakhst. 5 (1961) 374; Filim. in Opred. rast. Sr. Azii. [Key to Plants of Mid. Asia] 7 (1983) 355; Claves pl. Xinjiang. 3 (1985) 110.

Described from Tien Shan (Transili Ala Tau). Type in St.-Petersburg (LE).

On rocky slopes in midbelt of mountains.

IIA. **Junggar:** *Tien Shan* ("Ili"—Claves pl. Xinjiang. l.c.).

General distribution: Jung. Ala Tau, Nor. Tien Shan (Transili Ala Tau).

57. **O. taldycola** Grub. et Vass. in Novit. Syst. pl. vasc. 24 (1987) 138.

Described from Sinkiang (Tien Shan) Type in St.-Petersburg (LE).

In meadow slopes in midbelt of mountains?

IIA. **Junggar:** *Tien Shan* (Middle Taldy [river in Irenkhabirga mountain range], May 26, 1879—A. Reg., typus!).

General distribution: endemic.

Section 2. **Sphaeranthella** Gontsch.

58. **O. caespitosula** Gontsch. in Fl. URSS, 13 (1948) 551, 160; Fl. Kazakhst. 5 (1961) 391; Filim. in Opred. rast. Sr. Azii [Key to Plants of Mid. Asia] 7 (1983) 360; Claves pl. Xinjiang. 3 (1985) 100. —Ic.: Fl. SSSR, 13, Plate 7, fig. 3.

41 Described from Mid. Asia (West. Tien Shan). Type in St.-Petersburg (LE).

On rocky slopes, rocks, moraines in middle and upper belts of mountains.

IIA. **Junggar:** *Tien Shan* ("Baichen, Aksu"—Claves pl. Xinjiang. l.c.).

General distribution: Mid. Asia (West. Tien Shan).

Note. Could not locate Sinkiang specimens and, probably, this species was reported erroneously in Claves pl. Xinjiang. l.c. due to incorrect identification; it is regarded as endemic in West. Tien Shan.

59. **O. crassiuscula** Boriss. in Not. Syst. (Leningrad) 7, 11 (1937) 237; Vassilcz. and B. Fedtsch. in Fl. SSSR, 13 (1948) 163; Ikonnik. Opred. rast.

Pamira [Key to Plants of Pamir] (1963) 175; Filim. in Opred. rast. Sr. Azii [Key to Plants of Mid. Asia] 7 (1983) 362. —*O. microsphaera* auct. non Bge.: C.-y. Yang in Claves pl. Xinjiang. 3 (1985) 100. —**Ic.**: Boriss. l.c. fig. 2.

Described from Mid. Asia (Pamiro-Alay). Type in St.-Petersburg (LE).

On rocky slopes, talus, moraines, coastal pebble beds in upper and middle belts of mountains.

IB. Kashgar. *West.* ("south. slopes of Tien Shan"—Claves pl. Xinjiang. l.c.).
IIIC. Pamir ("West. Kunlun"—Claves pl. Xinjiang. l.c.).
General distribution: East. Pam.; Mid. Asia (Pam.-Alay).

Note. Closely related to *O. microsphaera* Bge. with a more western, west Tien Shan-Alay distribution range.

60. **O. valerii** Vass. in Novit. Syst. pl. vasc. 24 (1987) 133.

Described from Tibet (Weitzan). Type in St.-Petersburg (LE). Map 3.

On meadow slopes and coastal meadows in upper belt of mountains.

IIIB. Tibet: *Weitzan* (Konchyunchyu area, June 19 [July 1] 1884—Przew.; Russkoe lake and Yellow River bank [Huang He] 4100 m, June 17, 1900—Lad., typus!; Burkhan-Budda mountain range, nor. slope, Nomokhun gorge, 3650–3950 m, May 20, 1900; same site, south. and nor. slopes, Nomokhun gorge, 4100–4400 m, June 12, 1901; Yantszytszyan river basin, valley of Makhmukhchyu brook, 4100 m, May 21, 1901—Lad.).
General distribution: endemic.

Section 3. **Orobia** (Bge.) Aschers et Graebn.

61. **O. alpina** Bge. in Index sem. horti Dorpat. (1840) 8, No. 7; ej. Sp. Oxytr. (1874) 86; Saposhn. Mong. Alt. (1911) 363; Gr.-Grzh. Zap. Mong. [West. Mong.] 3, 2 (1930) 816; Kryl. Fl. Zap. Sib. [Flora of West. Siberia] 7 (1933) 1741; Vassilcz. and B. Fedtsch. in Fl. SSSR, 13 (1948) 86; Grub. Konsp. fl. MNR (1955) 187; Fl. Kazakhst. 5 (1961) 362; Ulzij. in Issl. fl. i rast. MNR [Study of the Flora and Plants of Mongolian People's Republic] 1 (1979) 124; id. in Grub. Opred. rast. Mong. [Key to Plants of Mongolia] (1982) 169; Opred. rast. Tuv. ASSR [Key to Plants of Tuva Autonomous Soviet Socialist Republic] (1984) 152; Claves pl. Xinjiang. 3 (1985) 107. —**Ic.**: Fl. Kazakhst. 5, Plate 51, fig. 2.

Described from Altay (upper Chui). Type in St.-Petersburg (LE).

On rocky slopes, talus, rocks and moraines, small coastal meadows and pebble beds, mossy-lichen rocky tundra in alpine belt and upper part of forest belt.

IA. Mongolia: *Khobd.* (east. slope of Sailyugem mountain range, Chyumali-Uttun river, June 14; Ulan-Daba pass, June 23; Tszusylan, July 11 and 13; south. tip of Kharkhira river, July 23—1879, Pot.), *Mong. Alt.* (Altyn-Cheche, June 22, 1870—Kalning; Taishiri-Ula, July 13 and 15, 1877—Pot.; Dain-Gol lake, June 27; Borborogusun river, Saksaya tributary, July 2, 1903—Gr.-Grzh.; Tsagan-Kobu road to Tsagan-Gol from Kalgutty, June 28; pass from Kak-Kul' lake to Ak-Kul' lake, July

17—1909, Saposhn.; upper Buyantu, Chigirtei-Gol beyond lake, nor. slope of Chigirtei-Ula, 2600–2800 m, July 4, 1971—Grub., Ulzij. et al.; Khasagtu-Khairkhan mountain range, nor. slope of Tsagan-Irmyk-Ula in upper Khunkerin-Ama, 3100 m, Aug. 23, 1972—Grub., Ulzij. et al.), *Gobi Alt.* (Ikhe-Bogdo-Ula, peak, rock screes, June 29, 1945—Yun.; Ikhe-Bogdo-Ula, plateau-like crest of mountain range, about 3700 m, July 29, 1948—Grub.), *Alash. Gobi* (Alashan mountain range, Yamata gorge, nor.-east. slope of upper and middle belts in forest, June 13, 1908—Czet.).

42

IIA. **Junggar:** *Cis-Alt.* (Kengeity river valley, Sept. 18, 1876—Pot.).

General distribution: West. Sib. (Altay), East. Sib. (Ang.-Sayans), Nor. Mong. (Fore Hubs., Hent., Hang.), China (Altay).

62. **O. altaica** (Pall.) Pers. Syn. pl. 2 (1807) 333; Bge. Sp. Oxytr. (1874) 19; Saposhn. Mong. Alt. (1911) 363; Gr.-Grzh. Zap. Mong. [West. Mongolia] 3, 2 (1930) 816; Kryl. Fl. Zap. Sib. [Flora of West. Siberia] 7 (1933) 1722; Vassilcz. and B. Fedtsch. in Fl. SSSR, 13 (1948) 88; Boriss. in Fl. Zabaik. [Flora of Transbaikalia] 6 (1954) 623; Grub. Konsp. fl. MNR (1955) 188; Fl. Kazakhst. 5 (1961) 362; Fl. Tsentr. Sib. [Flora of Cen. Siberia] 2 (1979) 611; Ulzij. in Issl. fl. i rast. MNR [Study of Flora and Vegetation of Mongolian People's Republic] 1 (1979) 125; id. in Grub. Opred. rast. Mong. [Key to plants of Mongolia] (1982) 169; Opred. rast. Tuv. ASSR [Key to Plants of Tuva Autonomous Soviet Socialist Republic] (1984) 152; Claves pl. Xinjiang. 3 (1985) 107. —*O. brevirostra* DC. Astrag. (1802) 64; ej. Prodr. 2 (1825) 277; Ledeb. Fl. Ross. 1 (1843) 590. —*Astragalus altaicus* Pall. Sp. Astrag. (1800) 56. —Ic.: Pall. l.c. tab. 45; DC. (1802) l.c. tab. 5 (sub nom. *O. brevirostra*).

Described from Altay. Type in St.-Petersburg (LE).

On small alpine meadows, rock screes, rocky-lichen tundra, sandy banks of brooks and lakes.

IA. **Mongolia:** *Mong. Alt.* (basin of Dain-Gol lake, June 27, 1903—Gr.-Grzh.; Karatyr river source, Aug. 3, 1908—Sap.; "Tyurgun', Dzhangyz-Agach, Urmogaity"—Saposhn. l.c.).

IIA. **Junggar:** *Cis-Alt.* ("Kran, alp."—Sap. l.c.).

General distribution: West. Sib. (Altay), East. Sib. (Ang.-Sayans), China (Altay).

63. **O. ambigua** (Pall.) DC. Astrag. (1802) 56; ej. Prodr. 2 (1825) 276; Bge. Sp. Oxytr. (1874) 76; Kryl. Fl. Zap. Sib. [Flora of West. Siberia] 7 (1933) 1733; Vassilcz. and B. Fedtsch. in Fl. SSSR, 13 (1948) 85; Grub. Konsp. fl. MNR (1955) 188; Fl. Kazakhst. 5 (1961) 361; Ulzij. in Issl. fl. i rast. MNR [Study of Flora and Plants of Mongolian People's Republic] 1 (1979) 124; id. in Grub. Opred. rast. Mong. [Key to Plants of Mongolia] (1982) 169; Opred. rast. Tuv. ASSR [Key to Plants of Tuva Autonomous Soviet Socialist Republic] (1984) 153; Claves pl. Xinjiang. 3 (1985) 107. — *O. uralensis* Lebeb. Fl. alt. 3 (1831) 289; ej. Fl. Ross. 1 (1843) 593 pro max. pte. —*Astragalus ambiguus* Pall. Sp. Astrag. (1800) 54. —Ic.: Pall. l.c. tab. 43, fig. A. and B; Fl. SSSR, 13, Plate 2, fig. fig. 3; Fl. Kazakhst. 5, Plate 1, fig. 4.

Described from Siberia. Type in London (BM).

On steppe slopes of mountains, larch forests and their borders, coastal meadows and scrubs.

IA. Mongolia: *Gobi Alt.* (Dundu-Saikhan mountain range, upper and middle belts, July 7, 1909—Czet.).
General distribution: Europe (Transvolga), West. Sib., East. Sib. (Sayans), Nor. Mong. (Fore Hubs., Hent., Hang.).

64. **O. argentata** (Pall.) Pers. Syn. pl. 2 (1807) 331; DC. Prodr. 2 (1825) 276; Bge. Sp. Oxytr. (1874) 102; Kryl. Fl. Zap. Sib. [Flora of West. Siberia] 7 (1933) 1747; Vassilcz. and B. Fedtsch. in Fl. SSSR, 13 (1948) 72; Fl. Kazakhst. 5 (1961) 358; Claves pl. Xinjiang. 3 (1985) 109. —*O. argyrophylla* Ledeb. Fl. alt. 3 (1831) 288. —*Astragalus argentatus* Pall. Sp. Astrag. (1800) 60. —**Ic.**: Pall. l.c. tab. 48; Ledeb. Ic. pl. fl. ross. (1829) tab. 54 (sub nom. *O. argyrophylla*).

Described from west. Altay. Type in London (BM).

On rocky steppe slopes of hills and ridges in foothills.

IIA. Junggar: *Cis-Alt.* ("along Chern. Irtysh valley"—Claves pl. Xinjiang. l.c.).
General distribution: West. Sib. (west. Altay, Irt.).

65. **O. chionophylla** Schrenk in Fisch. et Mey. Enum. pl. nov. 1 (1841) 77 and 2 (1842) 49; Ledeb. Fl. Ross. 1 (1843) 592; Bge. Sp. Oxytr. (1874) 87; Vassilcz. and B. Fedtsch. in Fl. SSSR, 13 (1948) 92; Grub. Konsp. fl. MNR (1955) 189; Fl. Kazakhst. 5 (1961) 363; Ulzij. in Issl. fl. i rast. MNR [Study of Flora and Vegetation of Mongolian People's Republic] 1 (1979) 125; id. in Grub. Opred. rast. Mong. [Key to Plants of Mongolia] (1982) 169; Filim. in Opred. rast. Sr. Azii [Key to Plants of Mid. Asia] 7 (1983) 348; Claves pl. Xinjiang. 3 (1985) 106. —**Ic.**: Fl. Kazakhst. 5, Plate 52, fig. 2.

43   Described from East. Kazakhstan (Jung. Ala Tau). Type in St.-Petersburg (LE). Map 3.

On rocky and rubbly steppe slopes, sedge-*Cobresia* wastelands, rocks, moraines, coastal pebble beds in upper belt of mountains.

IA. Mongolia: *Khobd.* (upper valley to Kalgutty [Dzhirgalantu] July 8, 1905—Sap.; Turgen' mountain range 3 km from Ogotor-Khamar-Daba pass, July 17; southwest. spur of Turgen' mountain range, nor. slope of Yamat-Ula mountains, 2700 m, July 19—1977, Sanczir et al.), *Mong. Alt.* (Taishiri-Ula, July 15, 1877—Pot.; Taishiri-Ula mountain range, south. slope in upper Shine-Usu, 2450 m, near crest, June 18; Kobdo river basin, Duro-Nur lake, Mansar-Daba pass on road to Delyun, 2760 m, June 30—1971, Grub., Ulzij., Dariima; Khasagtu-Khairkhan mountain range,, nor. slope of Tsagan-Irmyk-Ula, 3100 m, Aug. 23, 1972—Grub., Ulzij. et al.; Adzhi-Bogdo-Ula mountain range, plateau, 3000 m, Aug. 22, 1973—Isach. and Rachk.), *Gobi-Alt.* (Dundu-Saikhan mountain range, nor. slope, July 23, 1972—Guricheva and Rachk.).

IIA. Junggar. *Ala Tau* ("Bole, Ven'-tsyuan'"—Claves pl. Xinjiang. l.c.), *Tien Shan* (on nor.-east. fringe of Sairam lake, 2150-2450 m, Aug. 18; upper Borotala, 2600 m, Aug. 18, 1878—A. Reg.; upper Taldy, 2150-2450 m, May 16; upper Taldy, 2450 m, May 17; Taldy source, 3050 m, May 21; passage to Kash river from upper Taldy, 3050-3200 m, May 21; Taldy river, 2750 m, May 26; Taldy, 2450-2750 m, May 27; Irenkhabirga, Dumbedan near Kumbedan, 2450-2750 m, May 28; Kumbedan, 2750 m,

May 29; Kumbel', 2750–3050 m, May 31; Bagaduslung-Bainamun, east. tributary of Dzhin river, 1850 m, June 4; Naryngol on Tsagan-Usu, Dzhin canal, June 10; Chunkur-Daban between Ulyastai and Dzhungak, 3050 m, June 13; Karagol near pass to Nilki, 3050 m, June 17; Nilki brook, 1500–1850 m, June 19; Nilki brook near Kash, 2150 m, June 20; Nilki [between Ulyastai and Nilki], June 30; Borgaty pass, 2450–2750 m, July 7; Aryslyn, 2450 m, July 11; same site, 3050–3350 m, July 13; same site, 2750 m, July 17; Mengute, Irenkhabirga, 2750 m, Aug. 2—1879, A. Reg.).

General distribution: Jung.-Tarb.; (Jung. Ala Tau) Nor. Tien Shan (Transili Ala Tau).

66. **O. confusa** Bge. A. Lehmann reliq. bot. (1852) 77; ej. Sp. Oxytr. (1874) 79; Kryl. Fl. Zap. Sib. [Flora of West. Siberia] 7 (1933) 1735; Vassilcz. and B. Fedtsch. in Fl. SSSR, 13 (1948) 99; Fl. Kazakhst. 5 (1961) 366; Claves pl. Xinjiang. 3 (1985) 106. —Ic.: Fl. Kazakhst. 5, Plate 52, fig. 5. Described from Altay. Type in St.-Petersburg (LE).

On rocky steppe slopes of hills and mountains.

IIA. Junggar: Cis-Alt. ("basin of Chern. Irtysh river"—Claves pl. Xinjiang. l.c.). General distribution: West. Sib. (Altay, south.).

67. **O. frigida** Kar. et Kir. in Bull. Soc. natur. Moscou, 14 (1841) 402; Ledeb. Fl. Ross. 1 (1843) 593; Bge. Sp. Oxytr. (1874) 88; Kryl. Fl. Zap. Sib. [Flora of West. Siberia] 7 (1933) 1743; Vassilcz. and B. Fedtsch. in Fl. SSSR, 13 (1948) 94; Fl. Kazakhst. 5 (1961) 364; Filim. in Opred. rast. Sr. Azii [Key to Plants of Mid. Asia] 7 (1983) 348; Claves pl. Xinjiang. 3 (1985) 108. — Ic.: Fl. Kazakhst. 5, Plate 52, fig. 3.

Described from East. Kazakhstan (Tarbagatai mountain range). Type in St.-Petersburg (LE).

On rocky and rubbly steppe slopes in upper mountain belt.

IIA. Junggar: Tarb. (Hochebene Saur-Argultin, May 20-June 28, 1876—Pevtzow [Pevtsov]), Jung. Ala Tau (nor. slope of Kasan pass, 2750–3050 m, Aug. 11, 1878—A. Reg.), Tien Shan (upper Muzart valley below pass, 2750–3050 m, Aug. 19, 1877—Fet.; Mountains south-west of Ketmen' pass, 2750 m, June 22, 1878—A. Reg.; Yuldus, Sep. 1878—Fet.; Irenkhabirga, Taldy river source, 2450–3050 m, June 21–24; Aryslyn, Kash tributary, 3050–3350 m, July 13; Dzhagastai, Yuldus source, 2750–2900 m, Sept. 6—1879, A. Reg.; prope flumen Tekes, 1886—Krassnow [Krasnov]; Kotyl' pass on road to Yuldus from Karashar, 3100 m, Aug. 15, 1958—Yun. and I-f. Yuan').

General distribution: Jung.-Tarb., Nor. Tien Shan ?; West. Sib. (Altay, south-west.).

68. **O. gebleri** Fisch. ex Bge. in Mem. Ac. Sci. St.-Petersb. 22, 1 (1874) 89; Gr.-Grzh. Zap. Mong. [West. Mong.] 3, 2 (1930) 816; Vassilcz. and B. Fedtsch. in Fl. SSSR, 13 (1948) 81; Grub. Konsp. fl. MNR (1955) 189; Ulzij. in Issl. fl. i rast. MNR [Study of Flora and Vegetation of Mongolian People's Republic] 1 (1979) 123; id. in Grub. Opred. rast. Mong. [Key to Plants of Mongolia] (1982) 168. —O. bungeana Schischk. in Kryl. Fl. Zap. Sib. [Flora of West. Siberia] 7 (1933) 1744.

Described from Altay. Type in St.-Petersburg (LE).

In moraines, sedge-*Cobresia* wastelands, small meadows, rocky slopes, rock screes in upper belt of mountains.

44   **IA. Mongolia:** *Mong.Alt.* (Koshagach steppe up to upper Kobdo river, spring and summer, 1897—Demidova; slopes of Shadzagain-Suburga pass, July 22, 1898—Klem.; Tiekty river [Terekty], July 6, 1903—Gr.-Grzh.; Tolbo-Kungei-Ala Tau, 3200 m, Aug. 5, 1945; Bus-Khairkhan mountain range, 3200 m, July 17, 1947; Kharagaitu-Daba pass in upper Indertiin-Gol, July 24, 1947—Yun.; nor. slope of Khan-Taishiri-Ul, larch forest 15 km south-east of Yusun-Bulak, Sep, 1, 1948—Grub.), *Gobi Alt.* (Ikhe-Bogdo mountain range, waterdivide of Narin-Khurimt-Ama and Ketsu-Ama, June 28, 1945—Yun.).
General distribution: West. Sib. (Altay), Nor. Mong. (Fore Hubs. west.; Hangay: Khan-Khukhei).

69. **O. grandiflora** (Pall.) DC. Astrag. (1802) 57; ej. Prodr. 2 (1825) 277; Ledeb. Fl. Ross. 1 (1843) 596; Turcz. Fl. baic.-dahur. 1 (1842) 296; Bge. Sp. Oxytr. (1874) 92; Kitag. Lin. Fl. Mansh. (1939) 291; Vassilcz. and B. Fedtsch. in Fl. SSSR, 13 (1948) 93; Boriss. in Fl. Zabaik. [Flora of Transbaikalia] 6 (1954) 618; Grub. Konsp. fl. MNR (1955) 190; Fl. Intramong. 3 (1977) 227; Fl. Tsentr. Sib. [Flora of Cen. Siberia] 2 (1979) 615; Ulzij. in Issl. fl. i rast. MNR [Study of Flora and Vegetation of Mongolian People's Republic] 1 (1979) 116; id. in Grub. Opred. rast. Mong. [Key to Plants of Mongolia] (1982) 167; Pl. vasc. Helanshan. (1986) 157. —*O. collina* Turcz. in Bull. Soc. natur. Moscou, 15 (1842) 741. —*Astragalus grandiflorus* Pall. Sp. Astrag. (1800) 57.—*Ic.:* Pall. l.c. tab. 46; Fl. SSSR, 13, Plate 2, fig. 4; Fl. Intramong. 3, tab. 115, fig. 1–8.

Described from Siberia (Transbaikal). Type in London (BM).

In forb and meadow steppes, steppe slopes of mountains and knolls.

IA. Mongolia: *Cen. Khalkha* (on mountain between Nalaikha and Gagtsa-Khuduk, June 24, 1841—Kirilov; Khavtagai-Tsagan hill 60 km south-east of Underkhan town, June 20, 1971—Dashnyam, Isach. et al.; Gurban-Zagal somon, Bor-Undur mountain, 1080 m, July 23, 1974—Golubkova and Tsogt), *East. Mong.* (between Abder river and Borol'dzhitu area, May 26—1889, Pot. and Sold., vicinity of Manchuria station, 1915—Nechaeva; "Shilingol distinct"—Fl. Intramong. l.c.).
General distribution: East. Sib. (Daur.), Nor. Mong. (Fore Hing., Mong.-Daur.), China (Dunbei, west).

70. **O. ketmenica** Saposhn. in Not. Syst. (Leningrad) 4 (1923) 134; Vassilcz. and B. Fedtsch. in Fl. SSSR, 13 (1948) 82; Fl. Kazakhst. 5 (1961) 361; Filim. in Opred. rast. Sr. Azii [Key to Plants of Mid. Asia] 7 (1983) 348; Claves pl. Xinjiang. 3 (1985) 108. —**Ic.:** Fl. Kazakhst. 5, Plate 51, fig. 3.

Described from Nor. Tien Shan (Ketmen' mountain range). Type in St.-Petersburg (LE).

On small alpine meadows and rubbly-melkozem slopes of high mountain areas.

IIA. Junggar: *Tien Shan* ("Chzhaosu, Chapchal"—Claves pl. Xinjiang. l.c.).
General distribution: Nor. Tien Shan (Ketmen' mountain range).

71. **O. latibracteata** Jurtz. in Not. Syst. (Leningrad): (1959) 369. —*O. strobilacea* Bge. in Mem. Ac. Sci. St.-Petersb. 22, 1 (1874) 103 ex pte., pro pl. chin.; Peter-Stib. in Acta Horti Gotob. 12 (1937) 77. —*O. strobilacea chinensis* Bge.: Gr.-Grzh. Zap. Kitai [West China] 3 (1907) 483. —*O. strobilacea* var. *chinensis* Bge. and var. *mongolica* Bge. in herb. —*O. uralensis* auct. non DC., nec Pall.: Ulbr. in Bot. Jahrb. 36, Beibl. 82 (1905) 66; id. in Feddes Repert. Beih. 12 (1922) 425.

Described from Qinghai (Nanshan). Type in St.-Petersburg (LE).

In alpine meadows and meadow slopes of mountains, 2750–4600 m alt.

IA. Mongolia: *Alash. Gobi* (Alashan mountain range, June 20–July 10, 1873—Przew.).

IIIA. Qinghai: *Nanshan* (Yusun-Khatyma river, 2750–3050 m, July 11 [23] 1880—Przew., typus!; Nanshan mountain range, mid-June 1873—Przew.; near Machan-Ula mountains, alp. belt, 3350–3650 m, July 23; same site, July 24 and 26—1879, Przew.; not far from Rako-Gol river, July 21, 1880—Przew.; nor. slope of Humboldt mountain range near Kuku-Usu area, 3650 m, June 10, 1894—Rob.; 86 km west of Xining, pass, Aug. 5, 1959—Peter.), *Amdo* (near Baga-Gorgi river, 2750 m, May 25, 1880—Przew.).

IIIB. Tibet: *Weitzan* (left bank of Dychyu river [Yangtzy], 3950–4650 m, June 25; mountain range between Talachyu and Bychyu rivers, 4400 m, July 7; south. bank of Orin-Nor lake, on hills, 4100–4400 m, July 30—1884, Przew.; Russkoe lake [Orin-Nor] and Yellow river [Huang He], 4100 m, on hills, June 26, 1900; Burkhan-Budda mountain range, nor. slope, Khatu gorge, 3350–3650 m, July 11, 1901—Lad.).

General distribution: Nor. Mong. (Hangay: Dzhargalantu river source), China (North, North-West).

45 **Note.** The report of this species on the south-eastern border of Hangay within the distribution range of closely-related *O. strobilacea* Bge. is somewhat doubtful. It is set off from the nearest point of the main range of *O. latibracteata* by 900 km northward under conditions that are atypical for this species (dune! in midbelt of mountains). Moreover, the differences between these species are minor and A. Bunge was perhaps right in regarding them as varieties of one species.

72. **O. longibracteata** Kar. et Kir. in Bull. Soc. natur. Moscou, 14 (1841) 403; Ledeb. Fl. Ross. 1 (1843) 594; Bge. Sp. Oxytr. (1874) 81; Kryl. Fl. Zap. Sib. [Flora of West. Siberia] 7 (1933) 1739; Vassilcz. and B. Fedtsch. in Fl. SSSR, 13 (1948) 76; Fl. Kazakhst. 5 (1961) Filim. in Opred. rast. Sr. Azii [Key to Plants of Mid. Asia] 7 (1983) 347; Opred. rast. Tuv. ASSR [Key to Plants of Tuva Autonomous Soviet Socialist Republic] (1984) 153; Claves pl. Xinjiang. 3 (1985) 108.

Described from Altay (Narym mountain range). Type in St.-Petersburg (LE).

Along floors of river valleys.

Distributed in border areas of Tarbagatai, south. Altay and East. Tannu-Ol.

General distribution: Jung.-Tarb.; West. Sib. (Altay), East. Sib. (Ang.-Sayans: East. Tannu-Ola).

73. **O. macrosema** Bge. in Mem. Ac. Sci. St.-Petersb. 22 1 (1874) 101; Kryl. Fl. Zap. Sib. 7 (1933) 1746; Vassilcz. and B. Fedtsch. in Fl. SSSR, 13 (1948) 79; Ulzij. in Issl. fl. i rast. MNR [Study of Flora and Vegetation of Mongolian People's Republic] 1 (1979) 123; id. in Grub. Opred. rast. Mong. [Key to Plants of Mongolia] (1982) 169; Opred. rast. Tuv. ASSR [Key to Plants of Tuva Autonomous Soviet Socialist Republic] (1984) 153.

Described from Altay (upper Chui river). Type in St.-Petersburg (LE).

In meadows, wastelands, rocky slopes, sparse larch forests in upper belt of mountains.

IA. **Mongolia:** *Khobd.* (Turgen' mountain range, Turgen'-Gola valley 7 km beyond estuary, right bank northerly slope, sparse larch forest, July 17; same site, forb-*Cobresia* thickets, July 17—1971, Grub., Ulzij., Dariima), *Mong. Alt.* (Bulgan river basin, Kharogaitu-Khutul' pass, alp. meadow and rubble screes, July 24, 1947—Yun.).

General distribution: West. Sib. (Altay).

74. **O. martjanovii** Kryl. in Acta Horti Petrop. 21 (1903) 6; Saposhn. Mong. Alt. (1911) 363; Kryl. Fl. Zap. Sib. [Flora of West. Siberia] 7 (1933) 1736; Vassilcz. and B. Fedtsch. in Fl. SSSR, 13 (1948) 80; Grub. Konsp. fl. MNR (1955) 191; Ulzij. in Issl. fl. i rast. MNR [Study of Flora and Vegetation of Mongolian People's Republic] 1 (1979) 123; id. in Grub. Opred. rast. Mong. [Key to Plants of Mongolia] (1982) 168; Opred. rast. Tuv. ASSR [Key to Plants of Tuva Autonomous Soviet Socialist Republic] (1984) 153; Claves pl. Xinjiang. 3 (1985) 109. —Ic.: Kryl. l.c. tab. 2.

Described from Altay (Chui mountain range). Type in St.-Petersburg (LE).

In montane steppes, rubbly and rocky slopes, solonetzic coastal meadows and sandy-pebble bed banks of rivulets, sparse larch forests.

IA. **Mongolia:** *Khobd.* (Katu river, right tributary of Bukon'-Bere [Bukhu-Muren], lower part of gorge, June 13; Altyn-Khatysyn area, June 17; Altyn-Khatysyn area on left bank of Bukon'-Bere river, June 18—1879, Pot.; "Khattu B.-M."—Saposhn. l.c.; Bukhu-Muren-Gol and Khub-Usu-Gol interfluve 7–8 km east-nor.-east of Bukhu-Muren somon, July 15, 1971—Grub., Ulzij., Dariima), *Mong. Alt.* (Terekty-Buyantu pass, July 14, 1906—Saposhn.; "Tsagan-Gol, Kak-Kul', Karaganty, Saksai, Delyun"—Saposhn. l.c.; upper Kharagaitu-Gol, left bank tributary of Bulgan-Gol, Aug. 24, 1947—Yun.; Buyantu river basin, Delyun area near Bukhu-Tumur cemetery, July 1; Buyantu river basin, Dzhangyz-Agach river near road-crossing to Kudzhurtu settlement, right bank, July 1; upper Bulgan-Gola, Kudzhurtu settlement, Artelin-Ama, July 3; upper Buyantu river, on mountains nor. of Chigirtei lake, beyond pass, 2500 m, July 5; Kobdo river basin, Sagsai-Gol near bridge along road to Khargantu-Gol, 1800 m, floodplain, July 6; Kobdo river 3 km beyond Ulan-Khusu, 1800 m, right bank, July 12—1971, Grub., Ulzij. Dariima).

General distribution: West. Sib. (Altay), Nor. Mong. (Hang.: Khan-Khukhei mountain range), China (Altay).

75. **O. recognita** Bge. in Mem. Ac. Sci. St.-Petersb. 22, 1 (1874) 79; Saposhn. Mong. Alt. (1911) 363; Kryl. Fl. Zap. Sib. [Flora of West. Siberia] 7 (1933) 1737; Vassilcz. and B. Fedtsch. in Fl. SSSR, 13 (1948) 62; Grub.

Konsp. fl. MNR (1955) 193; Fl. Kazakhst. 5 (1961) 357; Ulzij. in Issl. fl. i rast. MNR [Key to Flora and Vegetation of Mongolian People's Republic] 1 (1979) 122; id. in Grub. Opred. rast. Mong. [Key to Plants of Mongolia] (1982) 168; Filim. in Opred. rast. Sr. Azii [Key to Plants of Mid. Asia] 7 (1983) 347; Claves pl. Xinjiang. 3 (1985) 106. —Ic.: Fl. Kazakhst. 5, Plate 52, fig. 4.

Described from Altay (upper Chui river). Type in St.-Petersburg (LE).

On rubbly and rocky steppe slopes of mountains, sedge-*Cobresia* wastelands, sparse larch forests, moraines and mountain plateaus in upper and forest belts.

IA. Mongolia: *Mong. Alt.* (Kashagach steppe up to upper course of Kobdo river, spring and summer of 1897—Demidova; East. Sumdairyk river bank, between moraines, July 30, 1908—Saposhn.; "Tsagan-Gol, Kak-Kul', Aksu, Nizhn. lake of Kobdo, Onkattu, Karatyr, Dain-Gol alps"—Saposhn. l.c.; waterdivide of Buyantu river and Bulgan-Gol, Akhuntyn-Daba on Delyun-Kudzhurtu road, 3050 m, July 2; upper Buyantu river, Chigirtei-Gol 12 km beyond lake, Chigirtei-Ula, about 2800 m, July 4; upper Kobdo river, watershed of Khargantu-Gol and Donyagiin-Khara-Nur lake, on road to Dayan-Nur lake, 2550 m, July 7; Dayan-Nur lake, south. border of settlement, nor. slope of Yamatyn-Ul, 2350 m, July 10; Kobdo river 2 km beyond Zhilandy river estuary, right bank of Gurtyn-Ama creek valley, 2200–2300 m, July 11—1971, Grub., Ulzij., Dariima).

IIA. Junggar: *Cis-Alt.* ("Altai" [Shara-Sume]), *Tarb.* and *Jung. Ala Tau* ("from Dachen to Bole"—Claves pl. Xinjiang. l.c.).

General distribution: Jung.-Tarb.; Nor. and Cen. Tien Shan; West. Sib. (Altay), China (Altay).

76. O. soongorica (Pall.) DC. Astrag. (1802) 73; ej. Prodr. 2 (1825) 277; Ledeb. Fl. alt. 3 (1831) 287; ej Fl. Ross. 1 (1843) 595; Bge. Sp. Oxytr. (1874) 78; Kryl. Fl. Zap. Sib. [Flora of West. Siberia] 7 (1933) 1734; Vassilcz. and B. Fedtsch. in Fl. SSSR, 13 (1948) 98; Fl. Kazakhst. 5 (1961) 364; Filim. in Opred. rast. Sr. Azii [Key to Plants of Mid. Asia] 7 (1983) 348; Opred. rast. Tuv. ASSR [Key to Plants of Tuva Autonomous Soviet Socialist Republic] (1984) 152; Claves pl. Xinjiang. 3 (1985) 106. —*Astragalus soongoricus* Pall. Sp. Astrag. (1800) 63. —Ic.: Pall. l.c. tab. 51; Fl. SSSR, 13, Plate 3, fig. 2; Fl. Kazakhst. 5, Plate 52, fig. 1.

Described from Altay (west. foothills.). Type in London (BM).

In steppes, rubbly and rocky steppe slopes of knolls and low mountains, steppe meadows.

IA. Mongolia: *Mong. Alt.* (Bulgan river basin, upper Ulyastain-Gola, July 10; Arshantyn-Nuru, near Bilut-Ul, July 21, 1984—Kam. Dariima).

IIA. Junggar: *Cis-Alt.* ("Altay"—Claves pl. Xinjiang. l.c.), *Jung. Ala Tau* (Dzhair mountain range, Dzhair pass along road to Otu from Toli, Aug. 9, 1957—Yun. and I-f. Yuan'), *Tien Shan* (20 km south-east of Urumchi, 850 m, No. 6003, May 25, 1958—Lee and Chu (A.R. Lee (1959)), *Tarb.* (east. fringe of Saur mountain range, submontane trail along road to Altay from Karamai, July 4, 1959—Yun. and I-f. Yuan'; "Dachen"—Claves pl. Xinjiang. l.c.), *Zaisan* ? *Balkh.-Alak.* ?

General distribution: Fore Balkh., Jung.-Tarb.; West. Sib. (Irt., Altay), East. Sib. (Sayan).

77. **O. strobilacea** Bge. in Mem. Ac. Sci. St.-Petersb. 22, I (1874) 103 excl. pl. chin.; Kryl. fl. Zap. Sib. [Flora of West. Siberia] 7 (1933) 1747; Vassilcz. and B. Fedtsch. in Fl. SSSR, 13 (1948) 68; Boriss. in Fl. Zabaik. [Flora of Transbaikalia] 6 (1954) 620; Grub. Konsp. fl. MNR (1955) 193; Fl. Kazakhst. 5 (1961) 358; Fl. Tsentr. Sib. [Flora of Cen. Siberia] 2 (1979) 622; Ulzij. in Issl. fl. i rast. MNR [Study of Flora and Vegetation of Mongolian People's Republic] 1 (1979) 122 pro pte.; id. in Grub. Opred. rast. Mong. [Key to Plants of Mongolia] (1982) 169; Opred. rast. Tuv. ASSR [Key to Plants of Tuva Autonomous Soviet Socialist Republic] (1984) 154; Claves pl. Xinjiang. 3 (1985) 109. —*O. uralensis* auct. non DC.: Pavl. in Byull. Mosk. obshch. ispyt. prir., otd. biol. 38 (1929) 93. —**Ic.**: Fl. Zabaik. [Flora of Transbaikalia] 6, fig. 313; Fl. Kazakhst. 5, Plate 51, fig. 5.

Described from Altay (Chui mountain range). Type in St.-Petersburg (LE). Plate 2, fig. 4.

In larch forests and their borders, coastal scrubs, steppe and meadow slopes of mountains in upper and forest belts.

47 **IA. Mongolia:** *Khobd.* (in jugo Sailughem, June 21, 1869—Malewski), *Mong. Alt.* (Taishiri-Ula, July 15, 1877—Pot.; same site, 8 km south-east of Tszasaktu-Khana camp, Aug. 9, 1930—Pob.; Tsastu-Bogdo mountain range in upper Dzuilin-Gol, 3400 m, wasteland, June 24; upper Bulgan river, Ioltyn-Gol valley near Kudzhurtu settlement, Artelin-Sala creek valley, July 3; Dayan-Nur lake, nor. slope of Yamatyn-Ula, 2350–2500 m, July 9—1971, Grub., Ulzij., Dariima), *Gobi Alt.* (Dzun-Saikhan, tip of Elo creek valley, Aug. 23, 1931—Ik.-Gal.; Dundu- and Dzun-Saikhan mountain ranges, slopes of mountains and gorges, July–Aug. 1933—M. Simukova; Dzun-Saikhan, west. part, steppe slope in upper belt, June 19, 1945—Yun.).

General distribution: West. Sib. (Altay), East. Sib., Far East, Nor. Mong. (Fore Hubs., Hent., Hang.), China (Altay).

78. **O. tschujae** Bge. in Mem. Ac. Sci. St.-Petersb. 22, 1 (1874) 86; Kryl. Fl. Zap. Sib. [Flora of West. Siberia] 7 (1933) 1742; Vassilcz. and B. Fedtsch. in Fl. SSSR, 13 (1948) 82; Grub. Konsp. fl. MNR (1955) 194; Ulzij. in Issl. fl. i rast. MNR [Study of Flora and Vegetation of Mongolian People's Republic], 1 (1979) 124; id. in Grub. Opred. rast. Mong. [Key to Plants of Mongolia] (1982) 169; Opred. rast. Tuv. ASSR [Key to Plants of Tuva Autonomous Soviet Socialist Republic] (1984) 153. —**Ic.**: Grub. Opred... [Key..] Plate 88, fig. 404.

Described from Altay (Chui mountain range). Type in Paris (P).

On rubbly and rocky slopes, rocks, talus and moraines in montane steppe and alpine belts.

IA. Mongolia: *Mong. Alt.* (Von d. Stadt Ulassutaja bis zur Urotsch. Kosch-Agatsch, June 15—July 15, 1879—Pewzov [Pevtsov]; midcourse of Bor-Balgasun river, tributary of Saksai, July 2, 1903—Gr.-Grzh., upper Bulgan river near Kudzhurtu settlement, Artelin-Sala creek valley, July 3; upper Kobdo river, watershed of Khargantu-Gol and Donyagiin-Khara-Nur lake along road to Dayan-Nur, 2550 m, July 7—1971, Grub., Ulzij., Dariima).

General distribution: West. Sib. (Altay), East. Sib. (Tannu-Ola, East. Sayan).

Section 4. Ortholoma Bge.

79. **O. biloba** Sap. in Not. Syst. (Leningrad) 4, 17–18 (1923) 135; Vassilcz. and B. Fedtsch. in Fl. SSSR, 13 (1948) 141; Fl. Kazakhst. 5 (1961) 381; Filim. in Opred. rast. Sr. Azii [Key to Plants of Mid. Asia] 7 (1983) 359; Claves pl. Xinjiang. 3 (1985) 104.

Described from East. Kazakhstan (Saur mountain range). Type in St.-Petersburg (LE).

On rocky and rubbly slopes, pebble beds of rivers, alpine meadows in upper and middle belts of mountains.

IIA. **Junggar: Tarb.** ("Saur"—Claves pl. Xinjiang. l.c.), *Tien Shan* ("east. and cen. Urumchi region"—Claves pl. Xinjiang. l.c.).

General distribution: Jung.-Tarb. (Saur mountain range).

Note. The reference to the report of this species in Chinese Tien Shan arouses doubt and calls for verification.

80. **O. brachycarpa** Vass. in Not. Syst. (Leningrad) 20 (1960) 243; Vassilcz. and B. Fedtsch. in Fl. SSSR, 13 (1948) 139; descr. ross.; Fl. Kirgiz. 7 (1957) 384; Fl. Kazakhst. 5 (1961) 379; Ulzij. in Issl. fl. i rast. MNR [Study of Flora and Vegetation of Mongolian People's Republic] 1 (1979) 126; id. in Grub. Opred. rast. Mong. [Key to Plants of Mongolia] (1982) 168; Filim. in Opred. rast. Sr. Azii [Key to Plants of Mid. Asia] 7 (1983) 357. —*O. transversa* Vass. in Not. Syst. (Leningrad) 20 (1960) 244; Vassilcz. and B. Fedtsch. in Fl. SSSR, 13 (1948) 139, descr. ross. —*O. kashemiriana* auct. non. Cambess.: Bge. in Mem. Ac. Sci. St.-Petersb. 7 ser. 22, I (1874) 43 quoad pl. e Kurmekty. —Ic.: Fl. Kazakhst. 5, Plate 46, fig. 4.

Described from Nor. Tien Shan (Transili Ala Tau). Type in St.-Petersburg (LE).

On rocky and rubbly slopes, pebble beds of rivers, coastal meadows in middle belt of mountains.

IA. **Mongolia: Mong. Alt.** ("Khasagtu-Khairkhan mountain range Ulzij., l.c.).

Note. I did not see the cited specimen of this species from Mong. Altay and its identification is doubtful.

48   81. **O. dischroantha** Schrenk in Fisch. et Mey. Enum. pl. nov. (1841) 78; Ledeb. Fl. Ross. 1, 3 (1843) 587; Bge. Sp. Oxytr. (1874) 51; Vassilcz. and B. Fedtsch. in Fl. SSSR, 13 (1948); Fl. Kazakhst. 5 (1961) 386; Filim. in Opred. rast. Sr. Azii [Key to Plants of Mid. Asia] 7 (1983) 360; Claves pl. Xinjiang. 3 (1985) 102. —*O. algida* Bge. in Bull. Soc. natur. Moscou, 39, 2 (1866) 9; id. sp. Oxytr. (1874) 62. —Ic.: Fl. Kazakhst. 5, Plate 46, fig. 3.

Described from East. Kazakhstan (Jung. Ala Tau). Type in St.-Petersburg (LE). Plate III, fig. 3.

On stony and rocky slopes, forests and alpine meadows in upper and middle belts of mountains.

IIA. Junggar: *Jung. Ala. Tau* (south. slope of Jung. Ala Tau, Borotala river basin, before Koketau pass, July 21, 1909—Lipsky), *Tien Shan* [Sudliches Kiukonik Tal beim Lager unter Tschon Yailak Pass, June 15, 1908—Merzb.; "Cen. Tien Shan, Ili"—Claves pl. Xinjiang. l.c.).

General distribution: Jung.-Tarb.

Note. Claves pl. Xinjiang. l.c. reports this species from Qinhe (Chingil', Cis-Alt.) as well but is not entirely reliable. Peter-Stib. in Acta Horti Gotob. 12 (1937) 77 reports this species from Qinhei (J. Rock, No. 14490) but points to the absence of fruits on the specimens examined, thus rendering their identification unreliable.

82. **O. floribunda** (Pall.) DC. Astrag. (1802) 75; Ledeb. Fl. Ross. 1, 3 (1843) 586, excl. syn.; Bge. Sp. Oxytr. (1874) 56; Kryl. Fl. Zap. Sib. [Flora of West Siberia] 7 (1933) 1728; Vassilcz. and B. Fedtsch. in Fl. SSSR, 13 (1948) 142; Fl. Kazakhst. 5 (1961) 381; Filim. in Opred. rast. Sr. Azii [Key to Plants of Mid. Asia] 7 (1983) 357; Claves pl. Xinjiang. 3 (1985) 103. — *Astragalus floribundus* Pall. Sp. Astrag. (1800) 47. —Ic.: Pall. l.c. tab. 37; Fl. Kazakhst. 5, Plate 46, fig. 5.

Described from West. Siberia (Irtysh river). Type in St.-Petersburg (LE).

On rocky and sandy-rubbly steppe slopes of knolls and low mountains.

IIA. Junggar: *Cis-Alt.* ("Altay"—Claves pl. Xinjiang. l.c.), *Jung. Ala. Tau* (ascent to Kuzyun' pass, rocky site, Aug. 2, 1908—B. Fedtsch.), *Tien Shan* (between Chzhaosu and Shati [Syati] 2 km from highway, on lower part of trail, No. 842, Aug. 12, 1957—Shen Tyan).

General distribution: Fore Balkh., Jung.-Tarb., Nor. Tien Shan; Europe (Transvolga), West. Sib. (south.).

83. **O. fruticulosa** Bge. in Bull. Soc. natur. Moscou 39, 2 (1866) 7; ej. Sp. Oxytr. (1874) 49; Vassilcz. and B. Fedtsch. in Fl. SSSR, 13 (1948) 135; Fl. Kazakhst. 5 (1961) 378; Filim. in Opred. rast. Sr. Azii [Key to Plants of Mid. Asia] (1983) 339; Claves pl. Xinjiang. 3 (1985) 104.

Described form East. Kazakhstan (Jung. Ala Tau). Type in St.-Petersburg (LE), lost ?

On rocky slopes and talus, small alpine meadows in upper belt of mountains.

IIA. Junggar: *Cis-Alt.* ("Qinhe" [Chingil']—Claves pl. Xinjiang. l.c.), *Jung. Ala Tau* ("Bole" [Borotala]—Claves pl. Xinjiang. l.c.).

General distribution: Jung. Ala Tau.

Note. A. doubtful species with no specimen preserved either at St.-Petersburg or Alma Ata but its report in Jung. Ala Tau has been repeated by all taxonomists following its description by A. Bunge. The type of this species too could not be located. Only the Chinese investigators can explain the report to its occurrence in Sinkiang.

84. **O. grum-grashimailoi** Palib. in Bull. Herb. Boiss. 2 ser. 8, 3 (1908) 961; Gr.-Grzh. Zap. Mong. [West. Mongolia] 3, 2 (1930) 816; Vassilcz. and B. Fedtsch. in Fl. SSSR, 13 (1948) 146.—Ic.: Palib. l.c. tab. 4, fig. 5.

Described from Sinkiang (South. Altay). Type in St.-Petersburg (LE).
On rocky and stony steppe slopes.

49   IIA. Junggar: *Cis-Alt.* (south. slope of Chinese Altay, Kostuk river eastward of
estuary, June 8 [20], 1903—Gr.-Grzh., typus!).
General distribution: endemic ? (type specimen alone is known).

Note. Differs from closely related *O. hirsuta* Bge. only in the larger size of flowers
and broader leaflets numbering 3–5 (not 5–9) pairs.

85. **O. hirsuta** Bge. in Mem. Ac. Sci. St.-Petersb. 7 ser. 22, I (1874) 55;
Sap. Mong. Alt. (1911) 363; Kryl. Fl. Zap. Sib. [Flora of West. Siberia] 7
(1933) 1730; Vassilcz. and B. Fedtsch. in Fl. SSSR, 13 (1948) 146; Fl.
Kazakhst. 5 (1961) 384; Filim. in Opred. rast. Sr. Azii [Key to Plants of Mid.
Asia] 7 (1983) 357; Claves pl. Xinjiang. 3 (1985) 102.

Described from East. Kazakhstan (Zaisan lake). Type in St.-Petersburg
(LE). Plate III, fig. 5.
On rocky and stony steppe slopes of foothills and low mountains.

IIA. Junggar: *Cis-Alt.* (arid slopes between Kholostu and Chenkur, Burchum, Aug.
16, 1906—Sap,; "Qinhe, Fukhai, Burchum"—Claves pl. Xinjiang. l.c.),   *Tarb.*
("Khabakhe"—Claves pl. Xinjiang. l.c.).
General distribution: Fore Balkh. (Zaisan basin), Tarb.

86. **O. podoloba** Kar. et Kir in Bull. Soc. natur. Moscou, 15 (1842) 327;
Bge. Sp. Oxytr. (1874) 50; Vassilcz. and B. Fedtsch. in Fl. SSSR, 13 (1948)
144; Fl. Kazakhst. 5 (1961) 383; Filim. in Opred. rast. Sr. Azii [Key to
Plants of Mid. Asia] 7 (1983) 346; Claves pl. Xinjiang. 3 (1985) 103. —*O.
brachybotrys* Bge. in Mem. Ac. Sci. St.-Petersb. 7 ser. 22, I (1874) 53.

Described from East. Kazakhstan (Jung. Ala Tau). Type is St.-Petersburg
(LE).
On rubbly and rocky slopes, pebble beds of rivers in middle belt of
mountains.

IIA. Junggar: *Cis-Alt.* ("Qinhe" [Chingil']—Claves pl. Xinjiang. l.c.),   *Tien Shan*
(Burkhan-Tau [Tekes basin] June 5, 1878—Fet.; "west. and Cen. Tien Shan"—Claves
pl. Xinjiang. l.c.).
General distribution: Jung.-Tarb., Nor. Tien Shan.

Note. The report of the occurrence of this species in Cis-Alt. Junggar arouses doubt.

87. **O. pulvinoides** Vass. in Not. Syst. (Leningrad) 20 (1960) 245;
Vassilcz. and B. Fedtsch. in Fl. SSSR, 13 (1948) 147, descr. ross.; Fl.
Kazakhst. 5 (1961) 386; Filim. in Opred. rast. Sr. Azii [Key to Plants of Mid.
Asia] 7 (1983) 360; Claves pl. Xinjiang. 3 (1985) 102.

Described from East. Kazakhstan (Jung. Ala Tau). Type in Tashkent
(TAK).
On rocky and rubbly slopes in upper belt of mountains.

IIA. Junggar: *Cis-Alt.* ("Qinhe" [Chingil']—Claves pl. Xinjiang. l.c.),   *Tarb*
("Dachen"—ibid), *Tien Shan* ("Ili"—ibid).

60

General distribution: Jung.-Tarb.

Note. I did not see any specimens from Sinkiang.

88. **O. sarkandensis** Vass. in Fl. URSS, 13 (1948) 549 and 140; Fl. Kazakhst. 5 (1961) 379; Filim. in Opred. rast. Sr. Azii [Key to Plants of Mid. Asia] 7 (1983) 357; Claves pl. Xinjiang. 3 (1985) 103.

Described from East. Kazakhstan (Jung. Ala Tau). Type in St.-Petersburg (LE).

On montane slopes in forest belt.

IIA. **Junggar:** *Cis-Alt.* ("Qinhe"—Claves pl. Xinjiang. l.c.), *Tarb.* ("Dachen"—ibid).
General distribution: Jung.-Tarb.

Note. I did not see any specimens from Sinkiang.

50   89. **O. schrenkii** Trautv. in Bull. Soc. natur. Moscou, 33, 1 (1860) 486; Bge. Sp. Oxytr. (1874) 52; Vassilcz. and B. Fedtsch. in Fl. SSSR, 13 (1948) 141; Fl. Kazakhst. 5 (1961) 380; Filim. in Opred. rast. Sr. Azii [Key to Plants of Mid. Asia] 7 (1983) 358; Claves pl. Xinjiang. 3 (1985) 104. —*O. floribunda* var. *brachycarpa* Kar. et Kir. in Bull. Soc. natur. Moscou, 14 (1841) 402.

Described from East. Kazakhstan (Tarbagatai). Type in St.-Petersburg (LE).

On rocky and stony slopes in middle and upper belts of mountains.

IIA. **Junggar:** *Tarb.* ("Dachen" [Chuguchak]—Claves pl. Xinjiang. l.c.).
General distribution: Jung.-Tarb.

90. **O. tenuis** Palib. in Bull. Herb. Boiss., ser. 2, 8 (1908) 160; Gr.-Grzh. Zap. Mong. [West. Mong.] 3, 2 (1930) 817; Vassilcz. and B. Fedtsch. in Fl. SSSR, 13 (1948) 144 in note; Ulzij. in Issl. fl. i rast. MNR [Study of Flora and Vegetation of Mongolian People's Republic] 1 (1979) 126; id. in Grub. Opred. rast. Mong. [Key to Plants of Mongolia]. (1982) 168. —**Ic.:** Palib. l.c. tab. 4, fig. 4.

Described from Sinkiang (Altay). Type in St.-Petersburg (LE).

On coastal solonetzic meadows in middle and upper belts of mountains.

IA. **Mongolia:** *Khobd.* (marshy interfluvine region of Bukhu-Muren and Khub-Usu-Gol 7–8 km east.-nor.-east of centre of Bukhu-Muren somon, solonetzic sedge-covered meadow, July 15, 1971—Grub., Ulzij. et al.), *Mong. Alt.* (Bugotor river—high Altay plateau between Burchum and Kran, June 11, 1903—Gr.-Grzh., typus!).
General distribution: endemic.

Note. Very close to *O. macrobotrys* Bge.

91. **O. tianschanica** Bge. in Mem. Ac. Sci. St.-Petersb. 7 ser. 14, 4 (1869) 43 and id. 22, 1 (1874) 49; Vassilcz. and B. Fedtsch. in Fl. SSSR, 13 (1948) 134; Fl. Kirgiz. 7 (1957) 383; Ikonnik. in Dokl. AN Tadzh. SSR, 20 (1957) 56; id. Opred. rast. Pamira [Key to Plants of Pamir] (1963) 173; Filim. in

Opred. rast. Sr. Azii [Key to Plants of Mid. Asia] 7 (1983) 358; Claves pl. Xinjiang. 3 (1985) 104. —*O. pulvinata* Saposhn. in Not. Syst. (Leningrad) 4, 17–18 (1923) 129; Vassilcz. and B. Fedtsch. in Fl. SSSR, 13 (1948) 14; Fl. Kirgiz. 7 (1957) 365; Claves pl. Xinjiang. 3 (1985) 89. —Ic.: Fl. SSSR, 13, Plate 5, fig. 4.

Described from Cen. Tien Shan. Type in St.-Petersburg (LE).

On rocky, rubbly and meadow slopes, rock screes and talus, moraines, coastal pebble beds in upper belt of mountains.

IB. **Kashgar:** *West.* (Muztag-Ata foothill, Subashi valley, July 20 and 23, 1909—Divn.; upper Kizylsu river beyond Kashgar, arid rocky slopes of Simkhan to Egin, July 1, 1929—Pop.; "south. slopes of Tien Shan up to Pamir"—Claves pl. Xinjinang. l.c.).

IIIC. **Pamir** ("Kongur mountain range, south-west. slope, left moraine of Kok-Sel' glacier, Yaman-Yarsu river, 4450 m, Aug. 8, 1956, Pen Shuli et al"—Ikonnik. l.c.).

General distribution: Cen. Tien Shan, East. Pam.; Mid. Asia (Pam. Alay).

Section 5. **Xerobia** Bge.

92. **O. ampullata** (Pall.) Pers. Syn. pl. 2 (1807) 332; DC. Prodr. 2 (1825) 278; Bge. Sp. Oxytr. (1874) 127; Kryl. Fl. Zap. Sib. 7 (1933) 1755; Vassilcz. and B. Fedtsch. in Fl. SSSR, 13 (1948) 181; Grub. Konsp. fl. MNR (1955) 188; Fl. Kazakhst. 5 (1961) 397; Ulzij. in Issl. fl. i rast. MNR [Study of Flora and Vegetation of Mongolian People's Republic] 1 (1979) 117; id. in Grub. Opred. rast. Mong. [Key to Plants of Mongolia] (1982) 167; Filim. in Opred. rast. Sr. Azii [Key to Plants of Mid. Asia] 7 (1983) 348; Opred. rast. Tuv. ASSR [Key to Plants of Tuva Autonomous Soviet Socialist Republic] (1984) 153; C.Y. Yang in Claves pl. Xinjiang. 3 (1985) 101. —*Astragalus ampullatus* Pall. Reise 3 (1776) 750. —Ic.: Fl. Kazakhst. 5, Plate 53, fig. 2.

Described from East. Siberia (Ang.-Sayan). Type in London (BM).

51   On rocky steppe slopes, rocks and talus, pebble bed steppes along river valleys.

IA. **Mongolia:** *Mong. Alt.* (Iter ad Chobdo, 1870—Kalning; on peak of Dolon-Nor-Daban mountain range, along granite rocks, July 8; on Taishiri-Ula mountain range beyond forest boundary, on rocks, July 15—1877, Pot.; Khan-Taishiri mountain range, south. macroslope in upper Shine-Usu, nor. of Khalyun somon, 2380–2450 m, on rocks in larch forest, June 18, 1971—Grub., Ulzij., Dariima); *Cen. Khalkha* (Ugei-Nur), *Gobi-Alt.* (Dzun-Saikhan mountain range, nor.-west. part, nor. slope, June 19, 1945—Yun.), *East. Gobi* (Del'ger-Hangay mountain range, nor. creek valley under peak, July 24, 1924—Pakhomov).

IIA. **Junggar:** *Cis-Alt.* ("Altay"—C.Y. Yang l.c.), *Tarb.* top of Kuzyun' pass, Aug. 2, 1908—B. Fedtsch.; "Saur"—C.Y. Yang l.c.), *Jung. Ala Tau* (Dzhair mountain range 1–1.5 km, nor.-east of Otu settlement along road to Chuguchak, steppe belt, along granite cone-shaped hills, Aug. 4; Maili mountain range, 40–42 km nor.-east of Junggar Inlet "meteorological station toward Karaganda pass," montane-steppe belt, on south. rocky slope, Aug. 14—1957, Yun. and I-f. Yuan'), *Tien Shan* (near Sochzhan river [nor. Karlyk-Taga foothill] June 12, 1877—Pot.; lateral creek valley of Taldy river,

2150–2750 m, May 15; upper Taldy, 2150–2750 m, May 15; upper Taldy, 2150–2450 m, May 16; upper Taldy, 2750–3050 m, May 20; Iren-Khabirga, Taldy, 2450 m, May 24–1879, A. Reg.).

General distribution: Fore Balkh., Jung.-Tarb., Nor. Tien Shan (Kungei-Ala Tau); West. Sib. (Altay), East Sib. (Ang.-Sayan), Nor. Mong. (Hent., Hang.).

93. **O. assiensis** Vass. in Not. syst. (Leningrad) 20 (1960) 246; id. in Fl. URSS, 13 (1948) 179, descr. ross.; Fl. Kirgiz. 7 (1957) 391; Fl. Kazakhst. 5 (1961) 397; Filim. in Opred. rast. Sr. Azii [Key to Plants of Mid. Asia] 7 (1983) 349; C.Y. Yang in Claves pl. Xinjiang. 3 (1985) 101. —Ic.: Fl. Kazakhst. 5, Plate 53, fig. 1.

Described from East. Tien Shan. Type in St.-Petersburg (LE).

On rubbly-rocky slopes, pebble bed valleys of rivers in middle and upper belts of mountains.

IIA. Junggar: *Tien Shan* ("Cen. Tien Shan, west. part"—C.Y. Yang l.c.).
General distribution: Jung.-Tarb. (Jung. Ala Tau), Nor. and Cen. Tien Shan.

94. **O. burchan-buddae** Grub. et Vass. in Novit. Syst. pl. Vasc. 24 (1987) 139.

Described from Tibet (Weitzan). Type in St.-Petersburg (LE).

On rubbly steppe slopes in alpine belt.

IIB. Tibet: *Weitzan* (Burkhan-Budda mountain range, Khatu gorge, nor. slope, 3350–3950 m, on clayey-rocky soil, July 11, 1901, No. 219—Lad., typus!).
General distribution: endemic.

95. **O. ciliata** Turcz. in Bull. Soc. natur. Moscou, 5 (1832) 186; Bge. Sp. Oxytr. (1874) 129; Fl. Intramong. 3 (1977) 225. —Ic.: Fl. Intramong. 3, tab. 112, fig. 1–6.

Described from Inner Mongolia (East. Mong.) Type in St.-Petersburg (LE).

On rubbly and rocky steppe slopes of mountains and knolls.

IA. Mongolia: *East. Mong.* (from Tsagan-Balgasu to nor.-west., on hill, 3 versts (1 verst = 1.067 km) afar, rocky soil, No. 2, May 1; around Tsagan-Balgasu, No. 20, May 4—1831, I. Kuznetsov, typus!; Kuitun, May 19, 1850—Tatarinov; South. Mongolia, around Kalgan, April 25 [May 5] 1871—Przew.; 20 km nor. of Khukh-Khoto town, through Datsinshan' pass, 1900 m, grassy-herb steppe, June 4, 1958—Petr.).
General distribution: China (North-West, North).

96. **O. diversifolia** Peter-Stib. in Acta Horti Gotob. 12 (1938) 78; Fl. Intramong. 3 (1977) 216, p. max. p.; Rast. pokrov Vn. Mong. [Vegetational Cover of Inner Mongolia] (1985) 149. —Ic.: Fl. Intramong. 3, tab. 109, fig. 7–11.

Described from Inner Mongolia (Suiyuan'). Type in Goteborg (GB).

On plains and dune sand, gorge floors in desert steppes and wastelands.

IA. **Mongolia:** *East. Mong.* ("Suiyuan-Liou, No. 2093, typus"—Peter-Stib. l.c.; "west. Shilingol' ajmaq [administrative territorial unit in Mongolia]"—Fl. Intramong. l.c.), *East. Gobi* ("Mongolia austr.—Hummel, No. 1081"—Peter-Stib. l.c.; "North of 52   Ulantsab ajmaq"—Fl. Intramong. l.c.), *Alash. Gobi* (east.—Rast. pokrov Vn. Mongolii [Vegetational Cover of Inner Mongolia] l.c.).

General distribution: endemic.

97. **O. eriocarpa** Bge. in Mem. Ac. Sci. St.-Petersb. 7 ser. 13, 1 (1874) 122; Saposhn. Mong. Altai (1911) 363; Kryl. Fl. Zap. Sib. [Flora of West. Siberia] 7 (1933) 175; Vassilcz. and B. Fedtsch. in Fl. SSSR, 13 (1948) 186; Grub. Konsp. fl. MNR (1955) 189; Fl. Kazakhst. 5 (1961) 400; Ulzij. in Issl. fl. i rastit. MNR [Study of Flora and Vegetation of Mongolian People's Republic] 1 (1979) 119; id. in Grub. Opred. rast. Mong. [Key to Plants of Mongolia] (1982) 168; Filim. in Opred. rast. Sr. Azii [Key to Plants of Mid. Asia] 7 (1983) 349; Opred. rast. Tuv. ASSR [Key to Plants of Tuva Autonomous Soviet Socialist Republic] (1984) 153; C.Y Yang in Claves pl. Xinjiang. 3 (1985) 100. —**Ic.:** Ulzij. l.c. (1979) 229, fig. 59.

Described from Altay. Type in St.-Petersburg (LE). Plate II, fig. 1.

On rubbly and rocky steppe slopes, thickets of *Cobresia* and cushion-shaped plants in montane steppe and alpine belts.

IA. **Mongolia:** *Khobd.* (Oigur river valley, July 15, 1909—Sap.; "Shar'yamaty"—Saposhn. l.c.), *Mong. Alt.* (Tsagan-Kobu road from Kalgutty to Tsagan-Gol, steppe slopes, June 28, 1905—Sap.; 'Tsagan-Gol, Kak-Kul'"—Saposhn. l.c.; Khadzhingiin-Nuru, Seterkhi-Khutul', 2800 m, pulvinoid steppe, June 25, 1971—Grub., Ulzij. et al.; Khasagtu-Khairkhan mountain range, Tsagan-Irmyk-Ula in upper Khunkeryn-Ama, 2700–3100 m, *Cobresia* thickets, Aug. 23, 1972—Grub., Ulzij. et al.).

IIA. **Junggar:** *Cis-Alt.* ("Altay"—C.Y. Yang l.c.), *Tarb.* ("Saur"—C.Y. Yang l.c.).

General distribution: Jung.-Tarb. (Tarb. ?); West. Sib. (Altay), East. Sib. (west. Sayans).

98. **O. intermedia** Bge. in Ind. sem. Horti Dorpat. 8 (1839), No. 13; ej. Sp. Oxytr. (1874) 123; Gr.-Grzh. Zap. Mong. [West. Mongolia] 3, 2 (1930) 816; Kryl. Fl. Zap. Sib. [Flora of West. Siberia] 7 (1933) 1754; Vassilcz. and B. Fedtsch. in Fl. SSSR, 13 (1948) 187; Grub. Konsp. fl. MNR (1955) 190; Ulzij. in Issl. fl. i rast. MNR [Study of Flora and Vegetation of Mongolian People's Republic] 1 (1979) 119; id. in Grub. Opred. rast. Mong. [Key to Plants of Mongolia] (1982) 168; Opred. rast. Tuv. ASSR [Key to Plants of Tuva Autonomous Soviet Socialist Republic] (1984) 153; C.Y. Yang in Claves pl. Xinjiang. 3 (1985) 100. —**Ic.:** Ulzij. l.c. (1979) 230, fig. 60.

Described from Altay. Type in St.-Petersburg (LE).

On rubbly and rocky steppe slopes of mountains and desert-steppe valleys of mountain rivers.

IA. **Mongolia:** *Mong. Alt.* (Dain-Gol lake basin, in mountains, 2290 m, June 22, 1903—Gr.-Grzh.; 12 km south-east of Khobdo town along road to Ulan-Bator, pass, steppe, Aug. 8, 1943—Luk'yanov; upper Kobdo river, Khoton-Gola valley, left bank, east. slope of Modot-Tologoi mountain, 2400 m, montane steppe, July 8, 1971—Grub., Ulzij. et al.), *Depr. Lakes* (Tes river [lower course] July 5, 1915—V. Tugarinova).

IIA. Junggar: *Cis-Alt.* ("East. Altay"—C.Y. Yang. l.c).
General distribution: West. Sib. (Altay), East. Sib. (Sayans).

99. O. junatovii Sancz. in Tr. inst. botan. AN MNR, 7 (1985) 90. —**Ic.:**
l.c. 91, fig. 1.
Described from Mongolia (Gobi-Alt.). Type in Ulan-Bator (UBA) [not UB
which is Acronym for Universidade de Brasilia, Brazil—Gen. ed.].
On arid- and desert-steppe rubbly and rocky slopes and trails of
mountains.

IA. Mongolia: *Gobi-Alt.* (nor. trail of Gurban-Saikhan mountain range along road
to Dalan-Dzadagad from Bayan-Dalai somon, feather grass desert steppe, May 6,
1941—Yun., typus!; Dzun-Saikhan mountain range, Khurmein somon, in lower and
middle mountain belts, June 19, 1945—Yun.).
General distribution: endemic.

100. O. klementzii Ulzij. in Bot. zh. 56, 12 (1971) 1795; Ulzij. in Issl. fl.
i rast. MNR [Study of Flora and Vegetation of Mongolian People's
Republic] 1 (1979) 121; id. in Grub. Opred. rast. Mong. [Key to Plants of
Mongolia] (1982) 168. —**Ic.:** Ulzij. l.c. (1979) 231, fig. 61.
Described from Mongolia (Hentey). Type in St.-Petersburg (LE).
On rubbly and rocky steppe slopes of mountains and knolls.

IA. Mongolia: *Cen. Khalkha* (Choiren-Ula—1940, S. Damdin; Bayan-Under
somon, Dagan-Del' somon along Ulan-Bator—Arbai-Khere road, rocky slope of knoll,
June 19, 1952—Davazamc; 20 km east-south-east of Dzhargalt-Khan, rocky slope of
knoll, petrophyte steppe, June 18; 26 km south-east of Under-Khan, Bayan-Khuduk-
Ula, 1580 m, petrophyte steppe, June 19—1971, Dashnyam, Isach. et al.; 60 km south
of Under-Khan, petrophyte steppe on knoll top, July 9, 1971—Isach. and Rachk.).
General distribution: Nor. Mong. (Hent., Hang.).

101. O. leptophylla (Pall.) DC. Astrag. (1802) 77; ej. Prodr. 2 (1825) 278;
Bge. Sp. Oxytr. (1874) 125; Danguy in Bull. Mus. nat. hist. natur. 8 (1913)
7; Peter-Stib. in Acta Horti Gotob. 12 (1938) 78; Vassilcz. and B. Fedtsch.
in Fl. SSSR, 13 (1948) 184; Boriss. in Fl. Zabaik. [Flora of Transbaikallia] 6
(1954) 615; Grub. Konsp. fl. MNR (1955) 191; Fl. Intramong. 3 (1977) 223;
Ulzij. in Issl. fl. i rast. MNR [Study of Flora and Plants of Mongolian
People's Republic] 1 (1979) 118; id. in Grub. Opred. rast. Mong. [Key to
Plants of Mongolia] (1982) 167. —*O. inschanica* H.C. Fu et Cheng f. in Fl.
Intramong. 3 (1977) 289. —*Astragalus leptophyllus* Pall. Reise, 3 (1776)
749.—**Ic.:** Pall. l.c. tab, 10; Fl. Zabaik. [Flora of Transbaikalla] 6, fig. 311; Fl.
Intramong. 3, tab. 113.
Described from East. Siberia (Dauria). Type in London (BM).
In petrophyte and sandy steppes, rocky steppe slopes of knolls.

IA. Mongolia: *Cen. Khalkha* (Gagtsa-Khuduk, June 23 (1841) Kirilov; "Vallee du
Keroulen, steppes [May 1896]—Chaff."—Danguy l.c.; 26 km south-east of Under-
Khan, Bayan-Khuduk-Ula, 1580 m, petrophyte steppe, June 19, 1971—Dashnyam,
Isach. et al.; 42 km west-nor.-west of Underkhan, rocky slope of knoll, sheep's fescue

steppe, June 18, 1971—Dashnyam, Isach, et al.), *East. Mong.* (in montosis Mongolia chinensis, 1831, Kuznetsov, Bge.; between Kulusutaevsk and Dolon-Nor, 1870—Lom.; Kalgan, Dec. 1871—Przew.; south-east. Mongolia, May 1872—Przew.; Kulun-Buirnur plain, Ongirtu lake, sandy steppe, May 28; same site, salt-deposited lake, among pea shrubs, May 29—1899, Pot. and Sold.; Manchuria railway station, steppe, June 6, 1902—Litw.; Zodol-Khan, steppe on basalts, May 14, 1944—Yun.; Nukhutyn-Daban pass 25 km south of Erdene-Tsagan somon among granite outliers, July 27, 1962—Dashnyam; Barun-Sul hill, petrophyte steppe at summit, July 1, 1971—Dashnyam, Isach. et al.).

General distribution: East. Sib. (Transbaikal), Nor. Mong. (Mong.-Daur.) China (Dunbei, North).

102. **O. micrantha** Bge. ex Maxim. in Bull. Ac. Sci. St.-Petersb. 26 (1880) 470; Gr.-Grzh. Zap. Mong. [West. Mongolia] 3, 2 (1930) 816; Grub. Konsp. fl. MNR (1955) 191; Ulzij. in Issl. fl. i rast. MNR [Study of Flora and Vegetation of Mongolian People's Republic] 1 (1979) 118; id. in Grub. Opred. rast. Mong. [Key to Plants of Mongolia] (1982) 167. —Ic.: Grub. Opred. rast. Mong. [Key to Plants of Mongolia] Plate 88, fig. 406.

Described from Nor. Mongolia (Hangay). Type in St.-Petersburg (LE).

In arid, sandy and montane steppes, rocky and rubbly slopes.

IA. **Mongolia**: *Khobd.* (Kharkyra group of hills, Burtu area, July 17, 1903—Gr.-Grzh.; 3–4 km west of Ulan-Daba pass, montane steppe, July 29, 1945—Yun.), *Mong. Alt.* (35–40 km south-south-east of Tugrik sume, Botkhyn-Ama area, steppe, Aug. 11, 1945—Yun.; Khasagtu-Khairkhan mountain range nor.-west of Chindamani-Ul along road to Daribi somon from centre of Gobi-Altay ajmaq [administrative territorial unit in Mongolia], rocky slope of creek valley, June 20; upper Bulgan-Gol, Ulagchin-Gola valley along road to Kudzhurtu 1 km before Khudzhirlag-Gola estuary, steppe, July 3—1971, Grub., Ulzij. et al.), *Depr. Lakes* (Dzapkhyn river 1 km before bridge on road to Gobi-Altay [Yusun-Bulak], desert steppe, Aug. 24, 1972—Grub., Ulzij. et al.), *Val. Lakes* (steppe between Mankhanei-Shinaga-Bulak and Khobur-Bulak Minor springs, June 30, 1894—Klem.; east. fringe of Guilin-Tal area, feather grass steppe, Aug. 26, 1943—Yun.).

General distribution: Nor. Mong. (Hang. west.).

103. **O. mixotriche** Bge. in Mem. Ac. Sci. St.-Petersb. Sav. Etrang. 2 (1835) 589; Turcz. Fl. baic.-dahur. 1 (1842) 300; Bge. Sp. Oxytr. (1874) 126; Vassilcz. and B. Fedtsch. in Fl. SSSR, 13 (1948) 183; Boriss. in Fl. Zabaik. [Flora of Transbaikallia] 6 (1954) 616; Grub. in Novosti sist. vyssh. rast. 9 (1972) 279; Ulzij. in Issl. fl. i rast. MNR [Study of Flora and Vegetation of Mongolian People's Republic] 1 (1979) 118; id. in Grub. Opred. rast. Mong. [Key to Plants of Mongolia] (1982) 168.

Described from East. Siberia (Daur.). Type in St.-Petersburg (LE).

On steppe slopes, fixed sand dunes.

IA. **Mongolia**: *Cen. Khalkha* (dunes near sources of Dzhargalant river, July 10, 1891, Levin.).

General distribution: East. Sib. (Daur.), Nor. Mong. (Hent., Hang.).

54   Note. A single locality of this species at the border of Cen. Khalkha and Hangay in Cen. Asia has been reported before: Grub. Konsp. fl. MNR (1955) 191 and Ulzij. l.c. (1979): 117 and 1982: 168 under related species *O. leucotricha* Turcz.

104. **O. monophylla** Grub. in Bot. zh. 63, 3 (1978) 364; Ulzij. in Bot. zh. 64, 9 (1979) 1230; id. in Grub. Opred. rast. Mong. [Key to Plants of Mongolia] (1982) 168. —*O. neimongolica* C.W. Chang et Y.Z. Zhao in Acta Phytotax. Sin. 19, 4 (1981) 523. —*O. diversifolia* auct. non Pet.-Stib.: Fl. Intramong. 3 (1977) 216 p. min. p. —**Ic.**: Bot. zh. 63, 3; 366; fig. 3; Acta Phytotax. Sin. 19, 4 : 524, fig. 1 (sub. *O. neimongolica*).

Described from Mongolia (East. Gobi). Type in St.-Petersburg (LE). Plate II, fig. 2. Map 2.

On Tertiary red earth and sand, along gorge floors, on rubbly and rocky slopes.

IA. **Mongolia:** *East. Mong.* ("Sonid Youqi" [Sunit]—Chang et Zhao l.c.), *Gobi Alt.* (30 km south of Bulgan somon centre, nor. trail of Gurban-Saikhan mountain range, along gorge floors, Aug. 10, 1972—Sanczir; 35 km south-east of Bayan-Dalai somon centre, intermontane basin, among chee grass thickets, Aug. 15, 1976—Gal.), *East. Gobi* (Bayan-Dzag area, eroded red sandstones, plateau and upper part of slopes, Oct. 21, 1947—Grub. and Kal., typus!; same site, precipices of brick-red Gobi formations, Sept. 5, 1950—Lavr. and Yun.; same site, basin floor, on rubbly trail, Sept. 13, 1979—Grub., Dariima, Muldashev; Shabarakh Usu, on sandy ridge at 3600 ft., 1925; Shara-Murun, near Ulu-Usu at 3700 ft., No. 25, 1925—Chaney), *Alash. Gobi* (Alashan mountain range, Tsuburgan-Gol gorge, April 26; same site, Tszosto gorge, nor.-west. and west. slopes, on sandy-rocky soil, May 10; same site, Tszosto gorge, south. slope, midbelt, on rubbly soil, May 15—1908, Czet.; environs of Nins' town, April 28, 1909—Napalkov: "Helanshan, Xiangchizigou, in declivibus lapidosis et apricis, alt. 2100 m, June 6, 1980—Zhao et Zhou", typus *O. neimongolica* !; Shilt-Ula south of Khurkhu mountain range, 120 km south-east of Nomgon somon centre, on outlier slope, July 15, 1974—Rachk. and Isach.).

General distribution: endemic.

Note. As noted on the labels, corolla is generally white.

105. **O. nitens** Turcz. in Bull. Soc. natur. Moscou, 15 (1842) 746; Bge. Sp. Oxytr. (1874) 93; Vassilcz. and B. Fedtsch. in Fl. SSSR, 13 (1948) 177; Boriss. in Fl. Zabaik. [Flora of Transbaikallia] 6 (1954) 614; Grub. Konsp. fl. MNR (1955) 192; Ulzij. in Issl. fl. i rast. MNR [Study of Flora and Vegetation of Mongolian People's Republic] 1 (1979) 115; id. in Grub. Opred. rast. Mong. [Key to Plants of Mongolia] (1982) 167.

Described from East. Siberia (Irkut river). Type in St.-Petersburg (LE).

On rubbly and rocky steppe slopes, sandy steppes and on fixed hummocky sand.

IA. **Mongolia:** *Cen. Khalkha* (west.: near Ugei-Nur lake, in loamy sand, steppe, July 14; same site, Boro-Undur mountain, along rocky slope, July 16—1924, Pavl.; sand along Dzhargalante river, June 1926—Zam,; waterdivide of Uber- and Ara-Dzhargalante rivers, hummocky sand overgrown with willow groves, July 2, 1949—Yun.).

General distribution: East. Sib. (Ang.-Sayans, Daur.), Nor. Mong. (Fore Hubs., Hent., Hang., Mong.-Daur.).

106. **O. potaninii** Bge. ex Palib. in Bull. Herb. Boiss. ser. 2, 8 (1908) 160; Saposhn. Mong. Altai (1911) 363; Gr.-Grzh. Zap. Mong. [West. Mongolia]

3, 2 (1930) 816; Grub. Konsp. fl. MNR (1955) 192; Ulzij. in Issl. fl. i rast. MNR [Study to Flora and Vegetation of Mongolian People's Republic] 1 (1979) 116; id. in Grub. Opred. rast. Mong. [Key to Plants of Mongolia] (1982) 167. —Ic.: Herb. Boiss, ser. 2, 8 tab. 3, fig. 2; Grub. Opred. rast. Mong. [Key to Plants of Mongolia] Plate 88, fig. 405.

Described from Mongolia (Mong. Alt.). Type in St.-Petersburg (LE).

On rubbly and rocky steppe slopes of mountains.

IA. Mongolia: *Mong. Alt.* ([Adzhi-Bogdo]? Bain-Gol, June 1877—Pot.; Koshagach steppe up to upper Kobdo river, spring and summer, 1897—Demidova; between Ikhes Nur and Tonkhil-Nur, on bank of arid bed, July 24, 1897—Klem.; Tiekty river [Terekty], July 6 [19] 1903—Gr.-Grzh., typus!; "Terekty river, barren steppe", July 14, 1906—Saposhn. l.c.; steppe around Kobdo town, July 18, 1906—Sap. sub. nom. *O. rhizantha* Palib.; upper Kobdo river, waterdivide of Khargantu-Gol and Donyagiin-Khara-Nur lake along road to Dayan-Nur [Dain-Gol], 2550 m, steppe, July 7, 1971—Grub., Ulzij. et al.; Ulan-Ergiin-Gol valley near centre of Must somon, granite knoll Mu-Ulan-Tologoi, gravelly-rocky slope exposed southward, Aug. 12, 1979—Grub., Dariima et al.).

General distribution: endemic.

107. **O. pseudofrigida** Saposhn. in Not. Syst. (Leningrad) 4 (1923) 136; Vassilcz. and B. Fedtsch. in Fl. SSSR, 13 (1948) 179; Fl. Kazakhst. 5 (1961) 396; Filim. in Opred. rast. Sr. Azii [Key to Plants of Mid. Asia] 7 (1983) 7 348; C.Y. Yang in Claves pl. Xinjiang. 3 (1985) 101.

Described from East. Kazakhstan (Jung. Ala Tau). Type in St.-Petersburg (LE).

On rocks, rocky slopes and short-grass meadows in upper belt of mountains.

IIA. Junggar: *Cis-Alt., Jung. Ala Tau, Tien Shan* ("from Chingil' river to Ili river"— C.Y. Yang l.c.).

General distribution: Jung.-Tarb., Nor. Tien Shan (Transili Ala Tau), Tien Shan (Terskei-Ala Tau).

108. **O. rhizantha** Palib. in Bull. Herb. Boiss., ser. 2, 8, 3 (1908) 159; Saposhn. Mong. Altai (1911) 363; Gr.-Grzh. Zap. Mong. [West. Mongolia] 3, 2 (1930) 817; Grub. Konsp. fl. MNR (1955) 193; Ulzij. in Issl. fl. i rast. MNR [Study of Flora and Vegetation of Mongolian People's Republic] 1 (1979) 120; id. in Grub. Opred. rast. Mong. [Key to Plants of Mongolia] (1982) 168. —Ic.: Palib. l.c. tab. 3, fig. 3.

Described from Mongolia (Mong. Alt.). Type in St.-Petersburg (LE).

On rubbly steppe slopes and trails of mountains, rocks, arid coastal pebble beds in montane steppe belt.

IA. Mongolia: *Khobd.* (Chyumalu-Uttun river, tributary of Kara-Mandai brook [east. slope of Sailyugem], along bank of arid pebble bed, June 13; Shar'-Gobu river valley, on dry slope of mountain, June 15; near Bairimen-Daban pass, south-west. slope of mountain, June 20—1879, Pot.; Koshagach steppe up to upper Kobdo river, spring and summer 1897—Demidova, paratypus!; beyond Tashenty pass, near Dandzhur-Nur lake, on mountain slope, July 1, 1898—Klem.), *Mong. Alt.* (between

Dain-Gol lake and Ak-Korum lake [about 2500 m] June 29 [July 12] 1903—Gr.-Grzh., typus!; "Kak-Kul' lake" [June 21], Kobdo town [July 14–19]—1906, Saposhn. l.c.; Duro-Nur lake, east. bank along road to Delyun, mountain trail, steppe, June 30; upper Buyantu river, Chigirtein-Gol 12 km beyond lake, right bank, terrace, 2350 m, steppe, July 4; upper Kobdo river, watershed of Khargantu-Gol and Dayangiin-Khara-Nur lake, on road to Dayan-Nur, 2550 m, hilly steppe, July 7—1971 Grub., Ulzij. et al.).

General distribution: endemic.

109. O. setifera Kom. in Feddes Repert. 13 (1914) 233.

Described from Sinkiang (Jung. Gobi). Type in St.-Petersburg (LE).

On arid mountain slopes ?

IIA. Junggar: *Jung. Gobi* (east.: Kuku-Syrkhe mountains, on sand formations (yellow-coloured) May 11 [23] 1879—Przew., typus!).

General distribution: endemic.

Note. V.L. Komarov in his description (l.c.) did not refer either to type specimen of this species or to its geographic distribution range or ecology. The Herbarium of the Komarov Botanical Institute of Russian Academy of Sciences has the solitary specimen cited by us. V.L. Komarov has himself written on the folder "*O. setifera*" and the specimen matches with the description, thus automatically rendering it the type.

110. O. setosa (Pall.) DC. Astrag. (1802) 56; Ledeb. Fl. alt. 3 (1831) 291; Bge. Sp. Oxytr. (1874) 121; Kryl. Fl. Zap. Sib. 7 (1933) 1751; Vassilcz. and B. Fedtsch. in Fl. SSSR, 13 (1948) 188; Grub. Konsp. fl. MNR (1955) 193; Ulzij. in Issl. fl. i rast. MNR [Study of Flora and Vegetation of Mongolian People's Republic] 1 (1979) 120; id. in Grub. Opred. rast. Mong. [Key to Plants of Mongolia] (1982) 168. —*O. aigulak* Saposhn. in Izv. Tomsk. otd. Russk. bot. obshch. 1 (1921) 30. —*O. irbis* Saposhn. ibid.—*Astragalus setosus* Pall. Sp. Astrag. (1800) 55. —Ic.: Pall. l.c. tab. 44 (sub nom. *Astragalus setosus*).

Described from Altay (Katun' river). Type in St. Petersburg (LE).

On rocky and rubbly arid slopes and rocks in montane steppe belt.

IA. Mongolia: *Khobd.* (in Jugo Sailughem, July 21, 1869—Malewski; nor.-east. slope of Ulan-Daban, lower forest boundary, June 22, 1879–Pot.).

General distribution: West. Sib. (Altay), Nor. Mong (Hang.: Khan-Khukhei mountain range).

111. O. stracheyana Benth. ex Baker in Hook. f. Fl. Brit. India, 2 (1876) 138; Hemsl. in J. Linn. Soc. (London) Bot. 35 (1902) 174; Fl. Xizang. 2 (1985) 870; Grub. in Novosti sist. vyssh. rast. 25 (1988) 108. —*O. poncinsii* Franch. in Bull. Mus. nat. hist. natur. 2 (1896) 343; O. and B. Fedtsch. Konspekt fl. Turkest. 2 (1909) 190; B. Fedtsch. in Fl. Tadzh. 5 (1937) 548; Persson in Bot. notiser (1938) 291; Vassilcz. and B. Fedtsch. in Fl. SSSR, 13 (1948) 178; Ikonnik. in Dokl. AN Tadzh. SSR, 20 (1957) 55; Fl. Kirgiz. 7 (1957) 390; Ikonnik. Opred. rast. Pamira [Key to Plants of Pamir] (1963) 171; Fl. Tadzh. 5 (1978) 490; Filim. in Opred. rast. Sr. Azii [Key to Plants of Mid. Asia] 7 (1983) 349; C.Y. Yang in Claves pl. Xinjiang. 3 (1985) 101. —

O. introflexa Freyn in Bull. Herb. Boiss. ser. 2, 6 (1906) 195. —Ic.: Fl. Xizang. 2, tab. 287, fig. 1–8.

Described from West. Tibet. Type in London (K). Isotype in St.-Petersburg (LE).

On rubbly, rocky and sandy-rocky slopes, moraines, clay ridges, pebble beds of rivers, sandy-pebble bed floors of gorges, alpine desert steppes and deserts.

IB. **Kashgar:** *West.* (Kok-Muinak pass along clayey descent, July 8; Subashi river valley, along rock screes, July 23—1909, Divn.; 3 km south-west of Torugart settlement on highway into USSR from Kashgar, alpine steppe, 3600 m, June 20, 1959—Yun. and I-f. Yuan').

IIA. **Qinghai:** *Nanshan* (Ritter and Humboldt mountain ranges, arid loessial descents of gorges, up to 3600 m, June 24; Sharagol'dzhin river, 3350 m, on sand, July 11; same site, Paidza-Tologoi area, 3350 m, sandy-pebbly steppe, July 11—1894, Rob.), *Amdo* (in upper Huang He, 2450–2750 m, on high pebble bed bank, May 19 [31] 1880—Przew.).

IIIB. **Tibet:** *Chang Tang* ("side of slope, 16,200 ft—1891, Thorold; 82°20', 35°20', 16,600 ft, July 27 [1896]—Wellby and Malcolm"—Hemsl. l.c.), *Weitzan* (left bank of Dychyu river [Yangtze] near Konchyunchyu estuary, on rocks, June 11 [23]; Alyk-Nor-Gol river valley, on sand, July 30 [Aug. 11]—1884, Przew.; Burkhan-Budda mountain range, south. slope, 4100 m, on arid clayey-rocky hill descents, June 2; Russkoe lake and Huang He river, 4100 m, on clayey ridges on banks of lake and river, June 14—1900; Russkoe lake, nor. bank, 4100 m, on rocks, along ridges on banks and on clay, June 2–3, 1901—Lad.).

IIIC. **Pamir** ("Tashkorghan, Dafdar, 3510 m, June 29, 1935"—Persson l.c.; valley of Sarykol river 10 km south of Bulunkul' settlement along road to Tashkurgan from Kashgar, pebble beds on central floodplain, June 12, 10–12 km south of Ulug-Rabat pass, along highway to Tashkurgan from Kashgar, 3900 m, alpine desert, June 14, 1959—Yun. and I-f. Yuan'; "Kungur, right moraine of Kok-Sel' moraine in upper Kenshirber-Su river, 4100 m, Aug. 16, 1956—Sh.-l. Pen and Skorobogatov"—Ikonnik. 1957 l.c.).

General distribution: Cen. Tien Shan, East. Pamir; Mid. Asia (Pamiro-Alay), Himalayas (Kashmir).

Section 6. **Baicalia** (Stell.) Bge.

112. **O. bicolor** Bge. Enum. pl. China bor. (1832) 17; ejusd. Sp. Oxytr. (1874) 150; Forbes and Hemsley, Index Fl. Sin. I (1887) 167; Peter-Stib. in Acta Horti Gotob. 12 (1938) 80; Fl. Intramong. 3 (1977) 234. —*O. uratensis* Franch. in Nouv. Arch. Mus. Paris. ser. 2, 5 (1883) 243; Forbes and Hemsley, Index Fl. Sin. 1 (1887) 167. —*O. angustifolia* Ulbr. in Bot. Jahrb. 36, Beibl. 82 (1905) 68. —Ic.: Fl. Intramong. 3, tab. 119, fig. 1–5.

Described from Nor. China (environs of Beijing). Type in Paris (P). Isotype in St.-Petersburg (LE).

57 On sandy banks of rivers and coastal dunes, sandy steppes and steppe slopes.

IA. **Mongolia:** *East. Mong.* (Mongolia chinensis, Bussun-tscholu, June 15 [27] 1850—Tatarinow; plain south of Kuku-Khoto, on sandy soil, July 6, 1884—Pot.; "Ourato, Tatsingchan, No. 2651, May 1866, David"—Franch. l.c. sub. nom. *O. uratensis*), *Ordos* (Echzhin-Khoro area, on clayey-sandy soil, Aug. 18, 1884—Pot.).

IIIA. **Qinghai:** *Nanshan* (San'chuan' area [basin along left bank of Huang He], 1850 m, June 7; valley of Dzhanby river [Itel'-Gol], June 13 and 15—1885, Pot.).

General distribution: China (North, North-West).

113. **O. chionobia** Bge. in Mem. Ac. Sci. St.-Petersb., 7 ser. 22, 1 (1874) 148 (Sp. Oxytr.); B. Schischk. in Fl. SSSR, 13 (1948) 204; Fl. Kirgiz. 7 (1957) 392; Fl. Kazakhst. 5 (1961) 405; Filim. in Opred. rast. Sr. Azii [Key to Plants of Mid. Asia] 7 (1983) 365; C.Y. Yang in Claves pl. Xinjiang. 3 (1985) 99. —*O. oligantha* auct., non Bge.: B. Schischk. in Fl. SSSR, 13 (1948) 205 quoad pl. songar.; Bajt. in Fl. Kazakhst. 5 (1961) 406 quoad pl. saur. —**Ic.:** Fl. SSSR, 13, Plate 11, fig. 3; Fl. Kazakhst. 5, Plate 53, fig. 3.

Described from East. Kazakhstan (Jung. Ala Tau). Type in St.-Petersburg (LE).

On rubbly and rocky slopes and rocks, talus and moraines, small alpine meadows, neve basins, from upper forest boundary to snow line.

IIA. **Junggar:** *Tarb.* (Saur mountain range, south. slope, valley of Karagaitu river, Bain-Tsagan creek valley, alpine belt, *Cobresia* meadow, June 23, 1957—Yun. and I-f. Yuan'), *Tien Shan* (Barkul'-Tag] on south. slope in forest belt, June 11; same site, close to upper forest boundary, June 11—1877, Pot.; south-west of Ketmen' pass, 2450-2750 m, June 20; Sumbe pass, 2750–3050 m, June 22; Koktyube mountains, 2150-2750 m, June 23—1878; Taldy, west. pass, 3350 m, May 20 (?); Iren-Khabirga, Taldy source, 3050 m, May 21 (?); Iren-Khabirga, Taldy river, 2750 m, May 26; Kumbel', 2750-3050 m, May 31; Kumbel', 3050 m, June 3; Chunkur-Daban nor. of Borborogusun, June 13; Chunkur-Daban between Shastai and Dzhumchaki, July 13—1879, A. Reg.).

IIIC. **Pamir** (moraine waterdivide between Atrakyr and Tyuzutek rivers, 4500-5000 m, mossy tundra July 20, 1942—Serp.).

General distribution: Jung.-Tarb., Nor. and Cen. Tien Shan; Mid. Asia (West. Tien Shan).

114. **O. fetissovii** Bge. in Acta Horti Petrop. 7 (1880) 367; B. Schischk. in Fl. SSSR, 13 (1948) 203; Fl. Kazakhst. 5 (1961) 404; Filim. in Opred. rast. Sr. Azii [Key to Plants of Mid. Asia] 7 (1983) 364; C.Y. Yang in Claves pl. Xinjiang. 3 (1985) 98. —*O. rigida* M. Pop. in Not. Syst. (Leningrad) 7 (1938) 116. —**Ic.:** Fl. SSSR, 13, Plate 10, fig. 1; Fl. Kazakhst. 5, Plate 50, fig. 4

Described from East. Kazakhstan (Jung. Ala Tau). Type in St.-Petersburg (LE).

On rubbly and clayey steppe slopes.

IIA. **Junggar:** *Cis-Alt.* (Qinhe region"—C.Y. Yang l.c.), *Tarb.* ("Dachen region"—C.Y. Yang l.c.).

General distribution: Jung.-Tarb.

115. **O. heterophylla** Bge. in Maxim. in Bull. Ac. Sci. St.-Petersb. 26 (1880) 480; Grub. Konsp. fl. MNR (1955) 190; Ulzij. in Issl. fl. i rast. MNR

[Study of Flora and Vegetation of Mongolian People's Republic, 1 (1979) 130; id. in Grub. Opred. rast. Mong. [Key to Plants of Mongolia] (1982) 166. Described from Sinkiang (Tien Shan). Type in St.-Petersburg (LE).

On rubbly, rocky and turf-covered steppe slopes, rocks and talus in alpine belt.

IA. Mongolia: *Khobd.* (mountains along left bank of Kharkhiry river, July 11, 1877; Ulan-Daba pass, June 22; south. tip of Kharkhira river, July 23; same site, on rocky mountain slopes, July 24—1879, Pot.; at Tashenty pass, July 1; beyond Tashenty pass, around Dandzhur lake, on mountain slope, July 1—1898, Klem.; Tsagan-Koby road from Kalgutta to Tsagan-Gol, steppe slopes, June 28, 1905—Sap.; Turgen' mountain range, Turgen'-Gol valley 7 km beyond estuary, right-bank slope, *Cobresia* herbage thickets, July 27, 1971—Grub., Ulzij. et al.; Mukhur-Ulyastyn-Gol gorge, forb steppe on south. slope, July 9; same site, slope with nor.-west. exposure, tundra, July 7—1977, Karamysheva, Sanczir et al.), *Mong. Alt.* (Altyn-Cheche lake, June 22, 1870—Kalning; Urmogaity pass, June 27; between Dain-Gol lake and Ak-Korum, June 29—1903, Gr.-Grzh.; Khara-Adzarga mountain range, Sakhir-Sala river valley, along east. rubbly slope of Imertsik mountain, Aug. 22; same site, around Khairkhan-Duru river, larch forest, Aug. 25–26 1930, Pob.; Adzhi-Bogdo mountain range, south. slope, Ikhe-Gol river, right flank of gorge, 2300–2500 m, on rocks, Aug. 22, 1979—Grub., Ulzij. et al.), *Depr. Lakes* (Dzun-Dzhargalantu mountain range, Ulyastyn-Gola gorge, 1850–2800 m, June 28, 1971—Grub., Ulzij. et al.).

IIA. Junggar: *Tien Shan* (declivitas meridionalis Thian-schan [ad trajectum] prope Hami, Aug. 31, 1875—Pias.; in prato planitiei altae, 2300–2750 m, Malyi Yuldus, June 3 [15] 1877—Przew., typus!; declivitas Thian-Schan [in valle fl. Tzanma] Juni 5 [17] 1877—Przew.), *Jung. Gobi* (nor.: slopes [of mountains] near Tsukhur-Nur lake, Sep. 20, 1930—Bar.).

General distribution: endemic.

116. **O. lanata** (Pall.) DC. Astrag. (1802) 89; Ledeb. Fl. Ross. 1 (1843) 581; Turcz. Fl. baic.-dahur. 1 (1842) 312; Bge. Sp. Oxytr. (1874) 145; Pavl. in Byull. Mosk. obshch. ispyt. prir., otd. biol. 38 (1929) 96; B. Schischk. in Fl. SSSR, 13 (1948) 197; Boriss. in Fl. Zabaik. [Flora of Transbaikalia] 6 (1954) 608; Grub. Konsp. fl. MNR (1955) 190; Ulzij. in Issl. fl. i rast. MNR [Study of Flora and vegetation of Mongolian People's Republic] 1 (1979) 131; id. in Grub. Opred. rast. Mong. [Key to Plants of Mongolia] (1982) 166. —*Phaca lanata* Pall. Reise, 3 (1776) 746. —Ic.: Pall. l.c. tab. A–a, fig. 2, 2B; Fl. Zabaik. [Flora of Transbaikalia] 6, fig. 308.

Described from East. Siberia (Daur.). Type in London (BM).

On hummocky sand and dunes, sandy banks of rivers and lakes.

IA. Mongolia: *Cen. Khalkha* (near Borokhchin lake, hummocky sand, July 6, 1924—Pavl.; Nalaikha, Elistyn-Daba, hummocky sand, July 14, 1924—Lis.; Ulan-Bator road—Tsetserleg, south. fringe of Tsagan-Nur lake, sand, June 25, 1948—Grub.; Dzun-Modo area 15 km east of Nalaikha town, sand, Sep. 7, 1949—Yun.).

General distribution: East. Sib. (Ang.-Sayan, Daur.), Nor. Mong. (Fore Hubs., Hang., Mong.-Daur.).

117. **O. langshanica** H.C. Fu in Fl. Intramong. ed. 2, 3 (1989) 672 and 313. —Ic.: l.c. tab. 121.

Described from Inner Mongolia. Type in Huhhot [Khukh-Khoto] (HIMC).

In sandy deserts.

IA. Mongolia: *Alash. Gobi* (Bayannolmeng, Wulatehouqi, Boyintu, June 7, 1982—Ge-ming, typus!).
General distribution: endemic.

118. **O. lanuginosa** Kom. in Feddes Repert. 13 (1914) 226; Ulzij. in Issl. fl. i rast. MNR [Study of Flora and Vegetation of Mongolian People's Republic] 1 (1979) 132; id. in Bot. zh. 64, 9 (1979) 1232; id. in Grub. Opred. rast. Mong. [Key to Plants of Mongolia] (1982) 166; Opred. rast. Tuv. ASSR [Key to Plants of Tuva Autonomous Soviet Socialist Republic] (1984) 151. —*O. pseudolanuginosa* Jurtz. in Novit. Syst. pl. vasc. (1964) 206.

Described from East. Siberia (Tuva) in St.-Petersburg (LE).

On steppe slopes and in sandy steppes.

IA. Mongolia: *Depr. Lakes* (right bank of Dzabkhyn river, on wet site, June 28, 1894—Klem.).
General distribution: East. Sib. (Tuva), Nor. Mong. (Hang. nor.-west.).

119. **O. lasiopoda** Bge. in Mem. Ac. Sci. St.-Petersb. 7 ser. 22, 1 (1874) 151; Pavl. in Byull. Mosk. obshch. ispyt. prir., otd. biol. 38 (1929) 95; Grub. Konsp. fl. MNR (1955) 191; id. in Novosti sist. vyssh. rast. 9 (1972) 293; Peshkova, Step. fl. Baikal. Sib. [Steppe Flora in Baikal Region of Siberia] (1972) 72; Ulzij. in Issl. fl. i rast. MNR [Study of Flora and Vegetation of Mongolian People's Republic] 1 (1979) 130; id. in Grub. Opred. rast. Mong. [Key to Plants of Mongolia] (1982) 166. —*O. elegans* Kom. in Feddes Repert. 13 (1914) 225; Grub. Konsp. fl. MNR (1955) 189; id. in Novosti sist. vyssh. rast. 9 (1972) 284.

Described from Mongolia (Cen. Khalkha). Type in St.-Petersburg (LE). Plate IV, fig. 4.

59   On rocky and rubbly steppe slopes of hills and foothills, solonetzic sandy steppes, coastal pebble beds and sand, solonetzic coastal meadows.

IA. Mongolia: *Cen. Khalkha* (Mongolia media prope Bussun-Tscholu, June 15 [27] 1850—Tatarinow, typus!; Gagtsa-Khuduk[1], June 23 [July 5] 1850—Tatarinow, paratypus!; Khara-Balgasun [Kara-Korum], sandy soil, June 20; near saline Qaidam lake, Aug. 3 1891, Levin; between Takhilta and Ugei-Nor, around salt lake, on solonchak, June 8, 1893—Klem., typus *O. elegans* Kom!; along east. bank of Ugei-Nor, June 11; same site, among iris thickets, June 17; in valley on right bank of Kukshin-Orkhon, June 18; not far from Khabyr-Nor lake, near well on sandy soil, June 19; in valley on right bank of Ongiin river, on fine rubble, July 27—1893, Klem.; 4 versts [1 verst = 1.067 km] from Chin-Tologoi mountains near well, among pea shrubs, May 19; between well around Chin-Tologoi mountain and Sudzha brook, in steppe on sandy soil, May 20; valley on right bank of Kharukhi river, between Khadasyn sentry post and Berkhe mountain, April 2—1894, Klem.; near Borokhchin lake, solonetzic sandy

---

[1]In Bunge's citation, erroneously recorded as "Gaschun-chuduk"! (Bge. l.c.).

steppe, July 6, 1924—Pavl.; Kholt area in Hangay foothills, June 6, 1926—Gus.; Ongiin-Gol, 1st rubbly terrace, July 18, 1926—Bulle; saline steppe near Dzaragyn-Gol brook south of Berkhe mountain range, Aug. 31, 1926—Prokh.; 7 km south of Bayan-Barat somon centre, on arid fringe of salt lake, June 13, 1951—Kal.), *East. Mong.* (Dabystu-Nor lake, on loose sandy soil, May 29–30; Elisyn-Khuduk well in Boro-Kholai area, on arid sandy soil, June 5; Khara-Tologoi hills, on rocky soil, June 7; steppe south of Kerulen river [Dolon-Dobo area], on sandy soil, June 10—1899, Pot. and Sold.), *Gobi Alt.* (Artsa-Bogdo, slopes near summit at 2300 m, No. 377, 1925—Chaney).

General distribution: East. Sib. (Daur.), Nor. Mong. (Hang., Mong.-Daur), China (Dunbei).

120. **O. mongolica** Kom. in Feddes Repert. 13 (1914) 226; Grub. Konsp. fl. MNR (1955) 191; Ulzij. in Issl. fl. i rast. MNR [Study of Flora and Vegetation of Mongolian People's Republic] 1 (1979) 131; id. in Grub. Opred. rast. Mong. [Key to Plants of Mongolia] (1982) 165; Opred. rast. Tuv. ASSR [Key to Plants of Tuva Autonomous Soviet Socialist Republic] (1984) 151. —Ic.: Grub. (1982) l.c., Plate 89, fig. 409.

Described from Mongolia (Depr. Lakes). Type in St.-Petersburg (LE). Map 2.

On coastal solonetzic sand and pebble beds.

IA. **Mongolia:** *Khobd.* (east. fringe of Achit-Nur lake basin, Ulyasutain-Gola valley, Aug. 1947—Tarasov), *Depr. Lakes* (on west. bank of Khirgis-Nur lake, on small loose coastal pebble bed, Aug. 1; Baga-Nur lake, Aug. 2; south. end of Khara-Usu lake, on arid solonetz, Aug. 18; Ubsu lake [opposite Baga-Nur], on pebble bed along banks, Sept. 22—1879, Pot.; from Ulyasutai to Kosh-Agach, June 15—July 15, 1880—Pevtsov; in valley on right bank of Bogden-Gol river, near mountain base, July 7, 1894; in left-bank valley of Bogden-Gol river, not far from Gudzhirtei area, on sandy hills, July 16, 1896—Klem. (lectotypus!); east. bank of Khara-Usu lake, and sand along road to Bayan-Khuduk, Aug. 12; fringe of Khara-Usu lake, chee grass, Aug. 14; Tuguryuk river, blown into chee grass thickets, Aug. 16; sand near Bayan-Khuduk, Aug. 19—1930, Bar.; 12 km south-east of Kobdo town along road to Ulan-Bator, summit of pass, grassy steppe, July 8, 1937—Luk'yanov).

General distribution: East. Sib. (cen. Tuva).

121. **O. myriophylla** (Pall). DC. Astrag. (1802) 87; Turcz. Fl. baic.-dahur. 1 (1842) 309; Bge. Sp. Oxytr. (1874) 141; Franch. pl. David. 1 (1884) 90; Pavl. in Byull. Mosk. obshch. ispyt. prir., otd. biol. 38 (1929) 96; Peter-Stib. in Acta Horti Gotob. 12 (1938) 81; Kitag. Lin. fl. Manshur. (1939) 292; B. Schischk. in Fl. SSSR, 13 (1948) 197; Boriss. in Fl. Zabaik. [Flora of Transbaikalia] 6 (1954) 608; Grub. Konsp. fl. MNR (1955) 191; Fl. Intramong. 3 (1977) 232; Ulzij. in Issl. fl. i rast. MNR [Study of Flora and Vegetation of Mongolian People's Republic] 1 (1979) 132; id. in Grub. Opred. rast. Mong. [Key to Plants of Mongolia] (1982) 165. —*O. davidii* Franch. Pl. David. 1 (1884) 89; Grub. and Ulzij. in Novosti sist. vyssh. rast. 27 (1990) 93 in synon.—*Phaca myriophylla* Pall. Reise, 3 Anhang (1776) 745. —*Astragalus myriphyllus* pall. Astrag. (1800) 87. —*A. verticillaris* L. Mant. (1771) 275. —Ic.: Pall. Reise, 3, tab. 7; Fl. Intramong. 3, tab. 118, fig. 9–15.

Described from East. Siberia (Daur). Type in London (BM). Paratype in St.-Petersburg (LE).

60   On sandy and meadow steppes, rubbly and rocky steppe slopes of hills and mud cones, river valleys.

IA. Mongolia: Cen. Khalkha (Tszakha mountains in Suchzh river valley, July 10, 1924—Pavl.; basin of Dzhargalante river, Uste mountain, nor. slope near summit, Aug. 12; Ubur-Dzhargalante river beyond breach near Botoga mountain, Sept. 2—1925, Krasch. and Zam.; environs of Ikhe-Tukhum-Nur lake, June 1926—Zam.; 26 km southeast of Under-Khan town, Bayan-Khutag mountains, 1580 m, June 19; Bayan-Khan-Ula 75 km nor.-west of Khutliin-Khuduk trail, July 15; Burgastyn-Ula mountains 20 km nor.-east of Bat-Norov somon, July 20—1971, Dashnyam, Karamysheva et al.; Munkh-Khan-Ula, 1607 m, July 25, 1974—Golubkova and Tsogt.), **East. Mong.** (in locis subarenosis Mongolia chinensis australis, 1831—Kuznetsov; Kuitun, May 19; Borol'dzhi, Dagkhan hills, June 17—1841, Kirilov; Mongolia chinensis, Schabartu, May 21–July 2, 1850—Tatarinov [Tatarinow]; Ourato, en montagne, June 1866, No. 2996—David, typus *O. davidii*!; south-east. Mongolia, Naryn-Shoron mountain [24–25] May 1872—Przew.; Angirtu lake, May 28; Abder river, June 25; between Abder brook and Borol'dzhi area, June 26; Ulan-Dzhilgu area, July 3—1899, Pot. and Sold.; Khailar railway station, June 10, 1902—Litw.; environs of Manchuria station, 1915—Nechaeva; Khoshun, Khalkhan-Gol, sand south of Khudugiin-Ukhane-Arag, June 9, 1925—Kazakevich; Khuntu somon, 17–20 km east-south-east of Bayan-Tsagan, Aug. 6, 1949—Yun.; Manchuria station, Nanshan mountain, 900 m, June 24, 1951—Ch. Wang, Sh.-s. Li et al.; Khailar town, Sishan' hills, 600 m, July 6, 1951—B. Skvortzov, A. Baranov et al.; Khamar-Daba region, June 18; Khalkha-Gol river floodplain in Khamar-Daba region, June 19; Khalkha-Gol river region, plain, June 19—1954, Dashnyam; Matad somon, 15 km east of Shorvog lake, raised sections of plain, June 5; Dege-Gol river region, south. slope of valley, June 10—1956, Dashnyam; Datsin'shan' mountain range, pass, 1900 m, June 4, 1958—Petr.; Shilin-Khoto town, steppe, 1959—Ivan.; 25 km south of experimental station near Khukh-Undur mountain [Khalkha-Gol somon], July 24; Zodol-Khairkhan mountains east of margin of Dariganga, July 28—1962, Dashnyam; 50 km nor. of Tamtsag-Bulak, June 27; Ottsog-Ula 67 km south of Badamyn-Khuduk, nor. slope, June 30; Barun-Sul hill, summit, July 1—1971, Dashnyam, Karamysheva et al.).

General distribution: East. Sib. (Ang.-Sayan—environs of Irkutsk, Daur.) Nor. Mong., China (Dunbei, North, North-West).

122. **O. ochrantha** Turcz. in Bull. Soc. natur. Moscou, 5 (1832) 188; Bge. Sp. Oxytr.. (1874) 149; Palib. in Acta Horti Petrop. 14, 1 (1895) 114; Peter-Stib. in Acta Horti Gotob. 12 (1938) 79; Kitag. Lin. Fl. Mansh. (1939) 292; Fl. Intramong. 3 (1977) 232; Fl. Xizang. 2 (1985) 870. —*O. chrysotricha* Franch. in Nouv. Arch. Mus. Paris, ser. 2, 5 (1883) 242. —*O. ochrocephala* var. *longibracteata* P.C. Li in Fl. Xizang. 2 (1985) 859. —**Ic.:** Fl. Intramong. 3, tab. 118, fig. 1–8; Fl. Xizang. 2, tab. 287, fig. 9–16.

Described from Inner Mongolia (East. Mong). Type in St.-Petersburg (LE).

On steppe slopes of mountains and flanks of gorges.

IA. Mongolia: East. Mong. (in locis subarenosis Mongoliae chinensis australis prope Tsagan-Balgassu, Majo 1831, leg. Kuznetsov-typus!; Mongolia chinensis,

1831—Bge.; Mongolia chinensis in reditu e Chinae, 1841—Kirilov; S.W. Mongolia, Muni-ula secus fl. Hoangho, med. July 1871, No. 207—Przew.; Oulachan, dans les vallees fraiches, No. 2812; "Ourato, lit desseche des torrents des montagnes centrales, No. 2859"—July 1866, David [sub nom. *O. chrysotricha* Franch.]; "Shilingol'sk ajmaq (administrative territorial unit in Mongolia), Datsin'shan', Ulashan' "—Fl. Intramong. l.c.).

IIIA. **Qinghai:** *Amdo* (in agris limosis secus rivulum Mudshik-che non procul a Huidul, 2750–2900 m, June 4 [1] 1880—Przew.).

General distribution: China (Dunbei west.; North, North-West).

Note. Corolla is usually light-yellow but keel often violet-coloured.

123. **O. oligantha** Bge. in Mem. Ac. Sci. St.-Petersb. Sav. Etrang. 2 (1835) 586; ej. Verz. suppl. fl. Alt. (1835) 85; Ledeb. Fl. Ross. 1 (1842) 582; Bge. Sp. Oxytr. (1874) 149; Sap. Mong. Alt. (1911) 364; Kryl. Fl. Zap. Sib. 7 (1933) 1763; B. Schischk. in Fl. SSSR, 13 (1948) 205; Grub. Konsp. fl. MNR (1955) 192; Fl. Kazakhst. 5 (1961) 406; Ulzij. in Issl. fl. i rast. MNR [Study of Flora and Vegetation of Mongolian People's Republic] 1 (1977) 133; id. in Grub. Opred. rast. Mong. [Key to Plants of Mongolia] (1982) 166; Opred. rast. Tuv. ASSR [Key to Plants of Tuva Autonomous Soviet Socialist Republic] (1984) 151; C.Y. Yang in Claves pl. Xinjiang. 3 (1985) 99. —*O. politovii* Sumn. in Not. syst. (Tomsk), 1–2 (1933); Kryl. Fl. Zap. Sib [Flora of West. Siberia] 7 (1933) 1762.

Described from West. Siberia (Altay). Type in St.-Petersburg (LE).

61 On rocks, rock screes and talus, in sheep's fescue-*Cobresia* wastelands and cushion plant beds in alpine belt, 2500–3700 m alt.

IA. **Mongolia:** *Khobd.* (along Suok river, June 16, 1870—Kalning; Bairimen-Daban summit, June 20, 1879—Pot.; Achit-Nur lake basin, upper Talain-Tologoi-Gola 3 km west of Kholbo-Nur lake along road to Tsagan-Nur from Tamozhen', July 14, 1971—Grub., Ulzij. et al.; 20 km west-south-west of Ulangom, Mukhur-Ulyasty-Gola gorge, tundra, July 7, 1971—Karamysheva, Sanczir et al.), *Mong. Alt.* (slope of Bol'shoi Ulan-Daban pass between Bodunchi and Narin rivers, July 19; way from Narin river to Senkul' river, July 24—1898, Klem.; Ak-Korum pass from Kulagash to Saksai, July 7, 1906—Saposhn.; "Tsagan-Gol, Kak-Kul', Tal-Nur, Bzau-Kul', Turgun', Alp."—Saposhn. l.c.; Urtu-Gol river valley, along nor. slope of mountain, Aug. 17, 1930—Pob.; Khan-Taishiri-Ula, 2600–2700 m, on steep east. slope, Aug. 10, 1945—Leont'ev; Duro-Nur lake, Mansar-Daba pass along road to Delyun, 2766 m, June 30; Akhuntyn-Daba on Delyun-Kudzhurtu road, 3050 m, July 2—1971, Grub., Ulzij. et al.; Khasagtu-Khairkhan mountain range, nor. slope of Tsagan-Irmyk-Ula in upper gorge of Khunkerin-Ama, about 3000 m, Aug. 23, 1972—Grub., Ulzij. et al.; Ikhe-Ulan-Daba pass on Must somon-Uenchi somon road, 2945 m, Aug. 13; Adzhi-Bogdo mountain range in upper Ikhe-Gola, watershed plateau, 3200–3300 m, Aug. 29—1979, Grub., Dariima et al.), *Depr. Lakes* (Dzun-Dzhargalantu mountain range, south-west. slope, Ulyastyn-Gola gorge, about 2800 m, June 28, 1971—Grub., Ulzij. et al.), *Gobi Alt.* (Ikhe-Bogdo mountain range: Narin-Khurimt-Ama creek valley in midportion of nor. slope, June 28; same site, in upper creek valley, plateau, June 28; same site, June 29; upper Ketsu-Ama creek valley, upper belt, June 29; flat crest of mountain range in upper Bityuten-Ama, June 29—1945, Yun.; Narin-Khurimt-Ama gorge, east. flank, 2900 m, July 28; plateau-like crest of mountain range, 3700 m, July 29—1948, Grub.; same site, south. slope, 3150 m, Aug. 4, 1973—Rachk. and Isach.).

IIA. **Junggar:** *Tarb.* ("Saur"—C.Y. Yang l.c.).
General distribution: Jung.-Tarb. (Saur); West. Sib. (Altay), Nor. Mong. (Hang., cen. and Khan-Khukhei).

124. **O. oxyphylla** (Pall.) DC. Astrag. (1802) 67; Ledeb. Fl. Ross. 1 (1848) 580; Turcz. Fl. baic-dahur. 1 (1842) 306, ex p.; Bge. Sp. Oxytr. (1874) 142; Kitag. Lin. Fl. Mansh. (1939) 292; B. Schischk. in Fl. SSSR, 13 (1948) 195; Boriss. in Fl. Zabaik. [Flora of Transbaikallia] 6 (1954) 612; Grub. Konsp. fl. MNR (1955) 192; Ulzij. in Issl. fl. i rast. MNR [Study of Flora and Vegetation of Mongolian People's Republic] 1 (1979) 127; id. in Grub. Opred. rast. Mong. [Key to Plants of Mongolia] (1982) 166; Opred. rast. Tuv. ASSR [Key to Plants of Tuva Autonomous Soviet Socialist Republic] (1984) 151. —*O. hailarensis* Kitag. in Tokyo Bot. Mag. 48 (1934) 907; Fl. Intramong. 3 (1977) 234. —*O. arenaria* Jurtz. in Novit. syst. pl. vascul. (1964) 210; Ulzij. in Issl. fl. i rast. MNR [Study of Flora and Plants of Mongolian People's Republic] 1 (1979) 128; id. in Grub. Opred. rast. Mong. [Key to Plants of Mongolia] (1982) 165. —*Astragalus oxyphyllus* Pall. Sp. Astrag. (1800) 90 excl. pl. alt. —Ic.: Pall. l.c. tab. 74; Fl. Zabaik. [Flora of Transbaikallia] 6, fig. 309; Kitag. l.c. fig. 23; Fl. Intramong. 3, tab. 120, fig. 1–8.

Described from East. Siberia (Daur.). Type in London (BM). Isotype in St.-Petersburg (LE).

On steppe slopes, arid and sandy steppes, sand banks of rivers and lakes, chee grass thickets, pine groves on sand.

IA. **Mongolia:** *Cen. Khalkha* (westernmost find: Gashun-Bulak area 5 km southeast of Choiren, Aug. 24, 1940; Bayan-Barat somon, 3–4 km nor. of Tsakhiryn-Dzhasa-Ula Aug. 10, 1962—Dashnyam; 11 km south-east of Underkhan, July 19, 1971—Dashnyam, Rachk. et al.; Undzhul-Ula 1850 m, July 3, 1974—Golubkova, Tsogt et al.), *East. Mong.*
General distribution: East. Sib. (Ang.-Sayan, Daur.), Nor. Mong., China (Dunbei).

Note. Gymnocarpous form is found all over the distribution range of this species and along with the hebecarpous form in the same population as noticed by this author on coastal sand of Buir-Nur lake. It is therefore not justified accept to gymnocarpous form as independent species *O. arenaria* Jurtz. Insofar as *O. hailarensis* Kitag. is concerned, the author of this species himself has later treated it as a synonym of *O. xyphylla* (Kitag. 1939 l.c.).

125. **O. pavlovii** B. Fedtsch. et Basil. in Bull. Soc. natur. Moscou, sect. biol. 38, 1–2 (1929) 96; Grub. Konsp. fl. MNR (1955) 192; Ulzij. in Issl. fl. i rast. MNR [Study of Flora and Vegetation of Mongolian People's Republic] 1 (1979) 129; id. in Grub. Opred. rast. Mong. [Key to Plants of Mongolia] (1982) 166. —**Ic.:** B. Fedtsch. et Basil. l.c. fig. 6.

Described from Mongolia (Val. Lakes). Type in Moscow (MW). Isotype in St.-Petersburg (LE). Map 4.

62 On rubbly and rocky steppe slopes and trails of mountains and knolls, rocks, sand and arid steppes.

IA. Mongolia: *Cen. Khalkha* (Dzhargalante river basin, rubbly ridges, Aug. 11, 1925—Krasch. and Zam.; Ikhe-Tukhum-Nur lake region, environs of Mishik-Gun monastery, July 11; Ikhe-Dzara mountains [80–85 km south-east of Choiren], Aug. 25—1926, Lis.; 5 km nor.-east of Choiren, small knolls, July 3, 1970—Banzragch, Karamysheva et al.; 5 km south of Deren somon centre, small knolls, June 14; 50 km from Mandal-Gobi, small knolls, June 14—1972, Rachk. and Guricheva), *Val. Lakes* (rubbly steppe in Tuin-Gol river valley, 732 m, Aug. 30, 1924—Pavl., typus! 60–80 km west of Dzag-Baidarik somon, Aug. 27, 1943; Khurz-Maral somon along road to Dzag-Baidarik somon from Baishintugarazh, Aug. 29, 1945; 40 km of Bayan-Khongor on Dzag-Tsagan-Olom road, July 11, 1947—Yun.; 8 km nor. of Nariin-Tel somon, Ulyasutai mountain, about 2000 m, July 30, 1952—Davazamc), *Gobi Alt.* (Ubten-Daban pass, Ikhe-Bogdo mountain range, Aug. 31, 1886—Pot.; Khalga pass between Barun-Saikhan and Dundu-Saikhan hills, Aug. 3; Bayan-Tsagan mountains, Aug. 10; same site, Khotun creek valley, Aug. 11; Dundu-Saikhan, hills Aug. 18; Dzun-Saikhan hills, Aug. 25—1931, Ik.-Gal.; Bayan-Tsagan mountains, July-Aug. 1933—Khurlat and Simukova; Dzun-Saikhan, commencement of trail close to road to Bayan-Dalai from Dalan-Dzadagad, July 22; pass between Dzun- and Dundu-Saikhan, along crest, July 22; same site, on flat ridge, July 22; same site, low belt of mountains, in dry valley, July 22—1943, Yun.; Dzun-Saikhan mountain range, middle and lower belts, June 19, 1945—Yun.; Ikhe-Bogdo mountain range, south-south-west. slope, Narin-Khurimt gorge near estuary, about 2500 m, July 30; east. slope of Khatsabchiin-Khara-Ula 20 km south-east of Bayan-Undur somon along road, on slope and along floor of gorge, Aug. 26—1948, Grub.; cen. part of Gurban-Saikhan ridge on road to Khushu-Khural, July 19; south. slope of Dundu-Saikhan mountain range 35 km south-west of Dalan-Dzadagad, July 20—1950, Kal.; Dzun-Saikhan mountain range, nor. slope, along road to Elo-Ama creek valley from Dalan-Dzadagad, near pass, about 2400 m, July 21, 1970—Grub., Ulzij. et al.; Ikhe-Bogdo-Ula, nor. macro slope, Ulyastei-Ama creek valley, along floor, July 1, 1971—Banzragch et al.; Arts-Bogdo mountain range, nor. slope, 2080 m, July 5; same site, summit, 2250 m, July 5—1973, Rachk. and Isach.; Gurban-Saikhan mountain range, Khabtsagaitu-Gol area, July 10, 1974—Rachk. and Volkova), *East. Gobi* (7 km south of Ikhe-Dzhargalant somon, plain, July 15; 35 km south-south-east of railway station No. 25, July 15—1971, Rachk. and Isach.).

General distribution: endemic.

126. **O. pellita** Bge. in Mem. Ac. Sci. St.-Petersb., 7 ser. 22, 1 (1874) 147; B. Schischk. in Fl. SSSR, 13 (1948) 205; Fl. Kazakhst. 5 (1961) 405; Jurtzev in Novosti sist. vyssh. rast. 1 (1964) 202; Filim. in Opred. rast. Sr. Azii [Key to Plants of Mid. Asia] 7 (1983) 365; C.Y. Yang in Claves pl. Xinjiang. 3 (1985) 99.

Described from East. Kazakhstan (Tarbagatai, Tastau mountains). Type in St.-Petersburg (LE).

On rubbly and rocky slopes, talus, around snow patches in upper mountain belt.

IIA. Junggar: *Tarb.* ("Saur"—Yang l.c.).

General distribution: Jung.-Tarb. (Tarb.); ? West. Sib. ("Altay"—Yang l.c.).

Note. The Herbarium of the Komarov Botanical Institute of Russian Academy of Sciences has only the the specimen cited by us. It was collected by Schrenk (see Jurtzev l.c.). with which some plants of *O. chionobia* Bge. are mixed up. The report to its occurrence in the latest works cited by us is evidently based on erroneous data of "Flora SSSR". C.Y. Yang (l.c.) cites it for Chinese South. Altay as well, which is highly doubtful. Could this species be simply an ecological form of *O. chionobia* plain?

127. **O. prostrata** (Pall.) DC. Astrag. (1802) 85; Turcz. Fl. baic.-dahur. 1 (1842) 309; Bge. Sp. Oxytr. (1874) 151; Danguy in Bull. Mus. nat. hist. natur. 19, 8 (1913) 7; B. Schischk. in Fl. SSSR, 13 (1948) 198; Boriss. in Fl. Zabaik. [Flora of Transbaikallia] 6 (1954) 607; Ulzij. in Issl. fl. i rast. MNR [Study of Flora and Vegetation of Mongolian People's Republic] 1 (1979) 131; id. in Grub. Opred. rast. Mong. [Key to Plants of Mongolia] (1982) 165. —*Phaca prostrata* Pall. Reise, 3, Anhang (1776) 744. —*Astragalus dahuricus* Pall. Sp. Astrag. (1800) 88. —Ic.: Pall. Reise, 3, tab. 10, fig. 2; ej. Sp. Astrag. tab. 72.

63 Described from Dauria (Tarei lake). Type in London (BM). Paratype in St.-Petersburg (LE).

On sandy-rubbly steppe slopes of knolls and valleys, solonetzic banks of lakes.

IA. Mongolia: *East. Mong.* ("Vallee du Keroulen, No. 1629, June 8, 1886, Chaff."— Danguy l.c.; 38 km east of Bayan-Terem somon, Yund-Zhabyir-Nur lake, slope of lake basin, June 21, 1971—Dashnyam, Karam. et al.).

General distribution: East. Sib. (Daur.), Nor. Mong. (Mong.-Daur.).

128. **O. przewalskii** Kom. in Feddes Repert. 13 (1914) 227; C.Y. Yang in Claves pl. Xinjiang. 3 (1985) 98. —*O. bogdoschanica* Jurts. [Jurtz.] in Novit. syst. pl. vasc. 1 (1964) 204.

Described from Sinkiang (East. Tien Shan). Type in St.-Petersburg (LE). On rocky slopes and talus in alpine belt.

IIA. Junggar: *Tien Shan* (Barkul'tag, south. slope [on road to Nan'shan'kou picket from Koshety-Daban pass], May 25 [June 6] 1879—Przew., typus! montes Bogdo-Ola et opp. Urumtschi, Lager am Sudrande des Bogdo-Ola, Aug. 26–29, 1908, No. 1347— Merzb., typus *O. bogdoschanica*; left bank of Manas river, Danu river valley 6–7 km beyond creek valley toward Danu pass, alpine belt, talus, July 22, 1957—Yun. and I-f. Yuan'; at Kalangou near Turfan, 2500 m, on shaded slope, No. 5837, June 24, 1958—Lee and Chu (A.R. Lee 1959)).

General distribution: endemic.

Note. The specimen collected by Merzbacher in Bogdo-Ula and described by B.A. Jurtzev (l.c.) as new species *O. bogdoschanica* Jurtz. represents a smaller form with no significant difference from *O. przewalskii* Kom.

129. **O. pumila** Fisch. ex DC. Prodr. 2 (1825) 279; Ledeb. Fl. Ross. 1 (1842) 582; Bge. Sp. Oxytr. (1874) 144; Saposhn. Mong. Alt. (1911) 364; Gr.-Grzh. Zap. Mong. [West. Mongolia] 3, 2 (1930) 817; Kryl. Fl. Zap. Sib. [Flora of West. Siberia] 7 (1933) 1761; B. Schischk. in Fl. SSSR, 13 (1948) 199; Grub. Konsp. fl. MNR (1955) 192; Hanelt and Davazamc in Feddes Repert. 70, 1–3 (1965) 42; Ulzij. in Issl. fl. i rast. MNR [Study of Flora and Vegetation of Mongolian People's Republic] 1 (1979) 129; id. in Grub. Opred. rast. Mong. [Key to Plants of Mongolia] (1982) 166; Opred. rast.

Tuv. ASSR [Key to Plants of Tuva Autonomous Soviet Socialist Republic] (1984) 151. —*O. inaria* Ledeb. Fl. Alt. 3 (1831) 273, non DC. —**Ic.**: Ledeb. Ic. pl. fl. ross. 5, tab. 457.

Described from West. Siberia (Altay). Type in St.-Petersburg (LE).

On desert-steppe rocky and rubbly slopes of mountains, rocks and talus, in sandy steppes.

IA. Mongolia: *Khobd.* (mountains between Uruktu and Kobdo rivers, July 2, 1898—Klem.; Khatu, Boku-Merin, Oigur, Suok"–Saposhn. l.c.)., *Mong. Alt.* (around a cliff of Bichigin-Nuru, July 18, 1896—Klem.; Terekty river, July 6, 1903—Gr.-Grzh.; Saksai river, steppe valley near Nikiforov, trading station, Aug. 1, 1909—Saposhn.; "Tsagan-Gol, Kakkul', Saksai, Delyun"—Saposhn l.c.; Khan-Taishiri mountain range, south. slope 2 km from pass along road, Sept. 3, 1948—Grub.; west. fringe of Tsetseg-Nur basin, Temetiin-Khukh-Ula, pass along road to Must somon, 2350 m, June 26; Buyantu river basin, west. spur of Bugu-Ula along road to Tashintu-Daba pass, 5 km north of pass, 2260 m, July 2—1971, Grub., Ulzij. et al.; 15 km south of Shine-Zhinst somon, south. slope of Tsagan-Khalgin-Tsakhir mountains, in gorge of foothill rocks, July 27, 1973—Rachk. and Isach.; plain 3 km south of Umne-Gobi somon, Aug. 3, 1977—Karamysheva, Sanczir et al.; Ushgiin-Barun-Kholai valley 6 km north of pass on road to Tonkhil somon-Tsetseg-Nur, Aug. 11; Ulan-Ergiin-Gol 3 km beyond centre of Must somon, 1900 m, floodplain, Aug. 12–1979, Grub., Dariima et al.), *Depr. Lakes* (on Altyn-Khadasu cliff between Shara-Bulak spring and Koshety brook, July 13, 1894—Klem.), *Gobi Alt.* ("Dundu-Saichan, Chalga, halbwustenartige Berghange, 2000 m, No. 1131, May 1962; Dund-Argalant, ca. 20 km SSO vom Buncagan-nur, Schotterfeld eines Trockentales, No. 2184, June 1962"—Hanelt and Davazamc l.c.; Ikhe-Bogdo-Ula, south. macroslope, 2950 m, Aug. 4; same site, 2700 m, rocky trail, 6—1973, Rachk. and Isach.).

General distribution: West. Sib. (south-east. Altay).

64

130. **O. ramosissima** Kom. in Feddes Repert. 13 (1914) 227; Fl. Intramong. 3 (1977) 220; Rast. pokrov Vn. Mong. [Vegetational Cover of Inner Mongolia] (1985) 151. —**Ic.**: Fl. Intramong. 3, tab. 3, fig. I-II.

Described from Inner Mongolia (Ordos). Type in St.-Petersburg (LE).

On shifting sand dunes and fixed sand and sand-covered slopes of mud cones in desert and desert-steppe zones.

IA. Mongolia: *East. Gobi* ("south. Ulantsaba"—Fl. Intramong. l.c.), *Alash. Gobi* ("Alashan Gobi"—Fl. Intramong. l.c.), *Ordos* (Baga-Gol river in Ordos, on loose sand of old dunes, Sept. 12, 1884—Pot., typus!; 25 km south-east of Otokachi town, dune sand, on gently inclined slope, Aug. 1; 5 km nor.-west of Ushinchi town, somewhat overgrown dune sand, Aug. 3; 30 km south of Dalatachi town near Chzhandanchzhao village, depression in dune sand, Aug. 10—1957, Petr.).

General distribution: China (North-West: Shanxi).

Note. This distinct species with procumbent branched stems and axillary 1–3-flowered inflorescences stands isolated in section Baicalia and should be treated as special subsection Ramosissima Grub. subsect. nova hoc loco (caulis ramosi, inflorescentiae axillares pauciflorae, calyx tubulosus vel campanulato-tubulosus).

131. **O. rhynchophysa** Schrenk in Bull. phys.-math. Ac. Petrop. 2 (1844) 196; Bge. Sp. Oxytr. (1874) 146; B. Schischk. in Fl. SSSR, 13 (1948) 202; Fl.

Kazakhst. 5 (1961) 402; Filim. in Opred. rast. Sr. Azii [Key to Plants of Mid. Asia] 7 (1983) 364; Claves pl. Xinjiang. 3 (1985) 97; Ulzij. and Gubanov in Byull. Mosk. obshch. ispyt. prir., otd. biol. 97, 5 (1987) 115. —*O. sumnewiczii* Kryl. in Animadv. Syst. Herb. Univ. Tomsk, 7–8 (1932); Kryl. Fl. Zap. Sib. [Flora of West. Siberia] 7 (1933) 1763. —Ic.: Fl. SSSR, 13, Plate 8, fig. 1 and Plate 10, fig. 2; Fl. Kazakhst. 5, Plate 53, fig. 5.

Described from Cen. Kazakhstan (Ulu Tau mountains). Type in St.-Petersburg (LE).

On rocky and rubbly steppe slopes of hills, knolls and valleys, rocks.

IA. **Mongolia:** *Mong. Alt.* ("Tsagan-Khutgol pass, Aug. 15, 1978—Ogureeva [MW]; Altan-Khukhiin-Nuru mountain range 45 km south of Umne-Gobi somon, July 30, 1986—Rotschild (MW)"—Ulzij., Gubanov, l.c.).

IIA. **Junggar:** *Zaisan* ("Burchum, Khabakhe [Kaba river], Zimunai"—Claves pl. Xinjiang. l.c.).

General distribution: Aralo-Casp., Fore Balkh.; West. Sib. (south-west. Altay).

132. **O. saurica** Sap. in Not Syst. (Leningrad), 4 (1923) 137; B. Schischk. in Fl. SSSR, 13 (1948) 206; Fl. Kazakhst. 5 (1961) 406; Filim. in Opred. rast. Sr. Azii [Key to Plants of Mid. Asia] 7 (1983) 365; C.Y. Yang. in Claves pl. Xinjiang. 3 (1985) 99.

Described from East. Kazakhstan (Saur). Type in St.-Petersburg (LE).

On rocky slopes and rocks, talus, snow patches, in upper mountain belt.

IIA. **Junggar:** *Tarb.* (Saur mountain range, south. slope of Karagaitu river valley, subalp. belt, along south. slope of mountains, June 23, 1957—Yun. and I-f. Yuan'; "Saur"—C.Y. Yang l.c.).

General distribution: Jung.-Tarb. (Saur).

133. **O. selengensis** Bge. in Mem. Ac. Sci. St.-Petersb., 7 ser., 22 (1874) 143; Pavl. in Byull. Mosk. obshch. ispyt. prir., otd. biol. 38, 1–2 (1929) 95; B. Schischk. in fl. SSSR, 13 (1948) 201; Boriss. in Fl. Zabaik. [Flora of Transbaikalia] 6 (1954) 610; Grub. Konsp. fl. MNR (1955) 193; Ulzij. in Issl. fl. i rastit. MNR [Study of Flora and Vegetation of Mongolian People's Republic] 1 (1979) 129; id. in Grub. Opred. rast. Mong. [Key to Plants of Mongolia] (1982) 166.

Described from East. Siberia (Daur.). Type in St.-Petersburg (LE).

In sandy, grassy and wormwood steppes, on rubbly steppe slopes of knolls.

65 IA. **Mongolia:** *Cen. Khalkha* (Khukhu-Khoshu-usu, July 24; Ikhe-Dzara hill, Aug. 25—1926, Lis.; environs of Ikhe-Tukhum-Nur, between Uber-Bulgan-Ama and Bulyastuin-Ama ravines, July 27; Tashigi and Dulga foothills, July 31; Kushosta mountain, Aug.; Urshelta mountain, Aug.—1926, Zam.; Undzhul somon, 1800 m, steppe, June 25; Boyasgalant hill nor. of Undzhul, June 28; upland south-west of Undzhul, June 29–1974, Golubkova and Tsogt), *East. Mong.* (in locis subarenosis Mongoliae chinensis, 1831, I. Kuznetsov; Bge. ibid, June 15, 1850—Tatarinov [Tatarinow]).

General distribution: East. Sib. (Daur.), Nor. Mong. (Hang., Mong. Daur.).

134. **O. sutaica** Ulzij. in Bot. zh. 64, 9 (1979) 1233; Ulzij. in Issl. fl. i rast. MNR [Study of Flora and Plants of Mongolian People's Republic] 1 (1979) 133; id. in Grub. Opred. rast. Mong. [Key to Plants of Mongolia] (1982) 166. Described from Mongolia (Mong. Altay). Type in St.-Petersburg (LE). Isotype in Ulan-Bator (UBA) [Not UB which is Acronym for Universidade de Brasilia, Brazil]. Plate III, fig. 4.

On talus around neve basins and in sedge-*Cobresia* wastelands.

IA. **Mongolia:** *Mong. Alt.* (Tsastu-Bogdo-Ula [Sutai-Ula], south-east. slope in upper Dzuilin-Gola, 3400 m, Sedge-*Cobresia* swampy wasteland, June 23, 1971— Grub., Ulzij. et al., typus!).

General distribution: endemic.

135. **O. viridiflava** Kom. in Feddes Repert. 13 (1914) 227; Grub. Konsp. fl. MNR (1955) 194; Hanelt and Davazamc in Feddes Repert. 70, 1–3 (1965) 42; Ulzij. in Issl. fl. i rast. MNR [Study of Flora and Plants of Mongolian People's Republic] 1 (1979) 129; id. in Grub. Opred. rast. Mong. [Key to Plants of Mongolia] (1982) 166. —*O. pumila* auct. non Fisch.: Pavl. in Byull. Mosk. obshch. ispyt. prir., otd. biol. 38 (1929) 95. — Ic.: Grub. Opred. rast. Mong. [Key to Plants of Mongolia] Plate 87, fig. 400.

Described from Mongolia (Hangay). Lectotype in St.-Petersburg (LE).

In sandy and rubbly low-grass steppes, slopes of knolls and in valleys, sandy and pebble bed banks of rivers, on rocks.

IA. **Mongolia:** *Cen. Khalkha* (Khara-Balgasun [Kara-Korum], July 20, 1891— Levin; around Ugei-Nor lake, June 9; same site, on east. bank, June 17; not far from Khabyr-Nor lake, near well, June 19; on ridge on right bank of Ongiin river, July 25; in valley on right bank of Ongiin river, July 27—1893, Klem.; between Takhilt river and Ugei-Nor lake, July 14, 1924—Pavl.; Ubur-Dzhargalant river, Aug. 11, 1925—Krasch. and Zam.; environs of Kholt region, slopes and ravines, July 13–16, 1926—Gus.; Tsagan-Sumein-Gol river bank, Aug. 25, 1926—Ik.-Gal.; between Erdeni-Dzu and Kukshin-Gol river, Aug. 28, 1926—Prokh.; 5 km west of Dzhargalt-Khan somon on road to Ulan-Bator, Aug. 31, 1949—Yun.; Kerulen valley 3 km beyond Undurkhan, meadow on right bank above floodplain, Aug. 9, 1989—Grub., Gub., Dariima), *Val. Lakes* (east. fringe of Guilin-Tala plain, Aug. 26, 1943—Yun.).

General distribution: Nor. Mong. (Fore Hubs., Hang,., Mong.-Daur.).

Section 7. **Gobicola** Bge.[1]

136. **O. gracillima** Bge. in Linnaea, 17 (1843) 5 in nota; ej. Sp. Oxytr. (1874) 160; Grub Konsp. fl. MNR (1955) 190; Ulzij. in Bot. zh. 64, 9 (1979)

---

65   [1]This section differs very little from preceding Baicalia Bge. and is related to it through intermediate species *O. selengensis* Bge. as acknowleged by its author (Bge. l.c. 1874; 160). Its characteristics (campanulate calyx, small flowers) are of little importance, relative and unstable and it would be logical to regard it as just a subsection of the former.

1233; id. in Grub. Opred. rast. Mong. [Key to Plants of Mongolia] (1982) 166; Fl. Intramong. ed. 2, 3 (1989) 309. —*O. psammocharis* Hance in J. Linn. Soc. (London), Bot. 13 (1873) 78; Forbes and Hemsl. Index Fl. Sin. 1 (1887) 168; Peter-Stib. in Acta Horti Gotob. 12 (1937) 81; Fl. Intramong. 3 (1977) 237. —*O. acutirostrata* Ulbr. in Bot. Jahrb. 36, Beibl. 82 (1905) 68. —*O. minutiflora* Jurtz. in Novit. Syst. pl. vasc. (1964) 207. —*O. oxyphylla* auct. non DC.: Maxim. Prim. fl. Amur. [Index fl. Pekin.] (1859) 470; Franch. Pl. David. 1 (1884) 90. —**Ic.**: Grub. Opred. rast. Mong. [Key to Plants of Mongolia] Plate 87, Fig. 399; Fl. Intramong. ed. 2, 3, tab. 120, fig. 1–5.

66   Described from East. Gobi. Type in Paris (P).

In sand steppes, fixed and semifixed sand, forests, along sandy-pebble bed and sand terraces and banks of rivers and lakes.

IA. Mongolia: *Cen. Khalkha* (nomerous reports in west. half of region—environs of Ugei-Nur lake, Borokhchin and Kharakhin-Gol sand, meander of Toly river, Dzhargalante river basin), *East. Mong.* ("Toumet; Ourato, No. 2601, No. 2710, June 1866, David"—Franch. l.c.; plain south of Kuku-Khoto, July 6; environs of Kuku-Khoto, Aug. 15; plain nor. of Khekou, Aug. 3; nor. bank of Huang He before Khekou, Aug. 4; Edzhin-Khoro area, Aug. 18; Ushkyun-Tukhum area—Aug. 29—1884, Pot.; environs of Ikhe-Bulak, Barun-Shire mountain, south. slope, Aug. 23, 1927—Zam.; Ongon-Elis sand, bank of Boro-Bulak spring, Sept. 13; Ongon-Elis sand, Sept. 17—1931, Pob.; Moltsog-Elis sand, May 16, 1944—Yun.; Khailar town, steppe, 1959—Ivan.: sand around Moltsog-Obo mountain 12 km south of Dariganga, July 6, 1971—Dashnyam, Isach. et al.), *Val. Lakes* (around Tatsin-Gol river, July 24, 1924—Pavl.), *East. Gobi* ("Mongolia media, deserto Gobi, 1841, Rosov"—Bge. l.c. typus!), *Ordos* (Ordos, Aug. 1871—Przew. paratypus!; around Ulan-Tologoi-Sume monastery, Sept. 6; Baga-Chikyr lake, Sept. 25, 1884—Pot.; 60 km west of Ushinchi town, along bank of Ulibu-Nor salt lake, Aug. 2; 35 km south-east of Khanginchi town, Aug. 7; Khantaichuan river valley 40 km south of Dalatachi town, Aug. 9; 24 km nor. of Dzhasakachi town, Aug. 15—1957, Petr.; Dzhasakachi town, flat summit of hillock, Aug. 15, 1958—Petr.).

General distribution: Nor. Mong. (Hang.: Khoitu-Tamir, Mong.-Daur.: Hangay-Daban west of Ulan-Bator), China (North, North-West: Shanxi, Shenxi).

137. **O. racemosa** Turcz. in Bull. Soc. natur. Moscou, 5 (1832) 187; Bge. Oxytr. (1874) 161; Ulzij. in Bot. zh. 64, 9 (1979) 1233; id. in Grub. Opred. rast. Mong. [Key to Plants of Mongolia] (1982) 166.

Described from East. Gobi. Type in St.-Petersburg (LE). Plate IV, fig. 2.

In arid rocky and desert steppes.

IA. Mongolia: *Khobd* (3 km west of Ulan-Daba pass, petrophyte steppe on top of knoll, July 15, 1977—Karam, Sanczir et al.), *Depr. Lakes* (Tes river, June 15, 1907—Dorogostaiskii), *East. Mong.* (Moltsog-Els sand 12 km south-east of Dariganga somon, June 16, 1980—Gub. [MW], *East. Gobi* (prope Chadatu, Aug. 1831—I. Kusnetsov, typus!; Kobur khuduk, June 25, 1841—Kirilov; "deserto Gobi prope Buchainmo-Ussu et Mogoitu", 1831—Bge. l.c.).

General distribution: China (North: Pohuashan mountains).

Section 8. Polyadena Bge.

138. **O. chiliophylla** Royle Ill. Bot. Himal. (1833) 198; Jacquem. Voy. Bot. (1844) 38; Bge. Sp. Oxytr. (1874) 155; B. Schischk. in Fl. SSSR, 13 (1948) 218; Fl. Tadzh. 5 (1978) 492; Filim. in Opred. rast. Sr. Azii [Key to Plants of Mid. Asia] 7 (1983) 365; Claves pl. Xinjiang. 3 (1985) 96; Fl. Xizang. 2 (1985) 869; Grub. and Ulzij. in Novosti sist. vyssh. rast. 27 (1990) 93. —*O. tibetica* Bge. Sp. Oxytr. (1874) 155; Keissler in Ann. Naturhist. Hofmus. 22 (1908) 24; Hand-Mazz. in Oesterr. bot. Z. 79 (1930) 33. —*O. grenardi* Franch. in Bull. Mus. nat. hist. natur. 3 (1897) 322. —*O. ingrata* Freyn in Bull. Herb. Boiss. Ser. 2, 6 (1906) 197. —*O. polyadenia* Freyn l.c. 199. —*O. physocarpa* auct. non Ledeb.: Hemsley in J. Linn. Soc. Bot. London 30 (1894) 111. —*O. microphylla* auct. non DC.: Henderson and Hume, Lahore to Jarkend (1873) 317; Hook. f. Fl. Brit. Ind. 2 (1876) 139; Alcock, Rep. natur. hist. results Pamir Boundary Commiss. (1898) 21; Deasy, in Tibet and Chin. Turk. (1901) 397; Hemsley, Fl. Tibet (1902) 174; Ostenf. and Pauls. in Hedin, S. Tibet, 6, 3 (1922) 65; Pamp. Fl. Carac. (1930) 152; Fl. Xizang. 2 (1985) 868; Claves pl. Xinjiang. (1985) 96. —**Ic.**: Jacquem. Voy. Bot. Tab. 45; Fl. Tadzh. 5 (1937) Plate 67.

Described from Kashmir. Type in London (K).

In alpine wormwood and winter fat rocky deserts, along arid pebble beds and sandy-pebble bed floors of river valleys, 2600–5500 m alt.

67 **IB. Kashgar:** *West.* (upper Tiznaf river 15 km beyond Kyude along Tibetan highway to Sarykdaban, short grass meadow along river bed, June 1, 1959—Yun. and I-f. Yuan'), *South.* (nor. slope of Russky mountain range along Keriya river, about 2600 m, April end to first half of May, 1885; Keriya mountain range, floor of gorge of Tulan-Khodzha river, on pebble bed, about 2600 m, May 19, 1885; Keriya mountain range, along Nura river, 2700 m, July 23, 1885—Przew.).

IIIB. Tibet: *Chang Tang* (Khotan province, Oitash down to the foot of Bushia glacier (northern side of Kuenluen), Aug. 27, 1856—Schlagintw.; Yarkend Exped., Drsub plains up to 5500 m, 1870—Henderson; Przewalsky mountain range, nor. slope 4200 m, on wet sandy soil, Aug. 20, 1890—Rob.; "Sandy gravelly soils in valleys at 5300 m, 1891—Thorold; 91°20′, 36°45′, June 1892—Rockhill; [Przewalsky mountain range] Aug. 1896—Hedin; 87°, 35° 18′, 5000 m, July 21, 1896—Wellby and Malcolm; near Mangtsa Tso, 5300 m, June 24 and 82°42′, 32°34′, 4500 m, Sept. 4—1896, Deasy and Pike"—Hemsley l.c.; Kar Yagde sur le Keria Daria, alt. 3910 m, Aug. 11, 1892, typus *O. grenardii!*; ibid alt. 4021 m, Aug. 12, 1892—D. de Rhins [P]; "Kan Jailok, 3180 m, June 19, 1906, Hedin"—Hand.-Mazz., l.c.; left bank of upper Karakash 10–12 km west of Shakhidulla along road to Kirgyz-daban, in valley of brook, on dry gravelly ridges, June 3, 1959—Yun. and I-f. Yuan'), *South.* (between Phari and Gyangtse, 3900–4500 m, June 19, 1904—Walton; "between Gyangchu-kamar, 4661 m and Tjarde, 4657 m, July 6; above the source of Tsangpo, northern foot of Himalayas, 5015 m, July 13—1907, Hedin"—Ostenf. and Pauls. l.c.).

IIIC. Pamir ("shore of Little Kara-Kul, 3720 m, July 15, 1894, Hedin"—Ostenf. and Pauls. l.c.; "in stony dry water-courses in Pamir region, at about 3900 m, No. 17702"—Alcock, l.c.; valley of nor. Geza, Mumi area, June 24, 1883—D. Schisch. ?; Muztag-ata foothill, along rock screes, July 20, 1909—Divn.; Kulin-aryk area between

Zad settlement and Tash-ui river, 3500–3800 m, July 29; sources of Kashkasu river, 4200–5500 m, July 5—1942. Serp.; 10–12 km south of Ulug-Rabat pass along road to Tashkurgan from Kashgar, wormwood-winter fat alpine desert, June 12; 42 km south of Bulunkul' up along Sarykol valley along road to Tashkurgan from Kashgar, 3900 m, alpine desert steppe, June 12; 10–12 km south of Ulug-Rabat pass, Sarykol' mountain range of highway, syrt (watershed upland) zone, 4050 m, alpine desert, June 14; west. fringe of Kongur peak at the point of penetration of Gez-dar'i 15 km away from Bulunkul' settlement along road to Kashgar from Tashkurgan, on gravel cone, June 15—1959, Yun. and I-f. Yuan').

General distribution: East. Pam.; Mid. Asia (Pamiro-Alay), Himalayas (Kashmir).

139. O. falcata Bge. in Mem. Ac. Sci. St.-Petersb., 7 ser. 22, 1 (1874) 156; Keissler in Ann. Naturhist. Hofmus. Wien. 22 (1907) 24; Peter-Stib. in Acta Horti Gotob. 12 (1937) 81; Hao in Bot. Jahrb. 68 (1938) 613; Grub. Konsp. fl. MNR (1955) 189; Ulzij. in Issl. fl. i rast. MNR [Study of Flora and Vegetation of Mongolian People's Republic] 1 (1979) 137; id. in Grub. Opred. rast. Mong. [Key to Plants of Mongolia] (1982) 167; C.Y. Yang in Claves pl. Xinjiang. 3 (1985) 97; Fl. Xizang. 2 (1985) 869; Grub. Novosti sist. vyssh. rast. (1988) 108. —O. holdereri Ulbr. in Notizbl. Bot. Gart. Berlin, 3 (1902) 193. —O. hedinii Ulbr. in Bot. Jahrb. 35 (1905) 680; Diels in Filchner, Wissensch. Ergebn. 10, 2 (1908) 254; Ostenf. and Pauls. in Hedin, S. Tibet, 6, 3 (1922) 64; Hand.-Mazz. in Oesterr. bot. Z. 79 (1930) 33; Ikonnik. Opred. rast. Pamira [Key to Plants of Pamir] (1963) 169; Fl. Tadzh. 5 (1978) 491; Filim. in Opred. rast. Sr. Azii [Key to Plants of Mid. Asia] 7 (1983) 366. —O. trichophysa auct. non Bge.: Batalin in Gr.-Grzh. Zap. Kitai [West China] 3 (1907) 483. —Ic.: Bot. Jahrb. 35 pl. 4, fig. 1; Ikonnik. l.c. Plate 20, fig. 5 (sub nom. O. hedinii).

Described from Qinghai (Nanshan). Type in St.-Petersburg (LE). Map 2.

In alpine steppes and deserts, on rubbly and rocky slopes, along floors of river valleys and gorges on sandy and pebble bed formations, 2500–5200 m alt.

IA. Mongolia: Mong. Alt. (Saksa river, on sandy soil, July 7, 1877—Pot.; 40 km nor. Tsogt somon, Tsakhir-Khalgana-Nuru, about 3100 m, on talus, Aug. 12, 1973—Isach. and Rachk.), Khobd. (Koshagach steppe up to upper Kobdo river, spring and summer 1897—S. Demidova).

IB. Kashgar: West. (Kun'-Lun', upper Tiznaf river 27 km beyond Kyude on road to Seryk-Daban pass, 3850 m, wormwood-winter fat desert on moraine, June 1, 1959—Yun and I-f. Yuan'), South. (Keriya mountain range, Ui-Eilak river bank, about 2500 m, on pebble bed, May 19, 1885—Przew.).

IC. Qaidam: montane (south. slope of South Kukunor mountain range, Dulan-Khit temple, 3050 m, May 8; Sarlyk-Ula mountains, Elisten-Kuku-Daban, 3350 m, May 22; Serik area, 3350 m, June 4; Ichegyn-Gol river, 3050 m, on sandy-pebbly bed, June 19—1895, Rob.).

IIIA. Qinghai: Nanshan (valley of Tetung river near Chortynton temple, early June 1873, No. 121—Przew., typus!; near Machan-Ula in alp. meadows, 3350–3650 m, July 24, 1879—Przew.; in desert nor. of Yuzhno-Kukunor mountain range, 3050 m, along brooks, July 5, 1880—Przew.; valley of Nan'chuan' river [around Xining town], May 2,

1885—Pot.; high mountains between Khsan and Tashitu rivers, May 25, 1886—Pot.; Xining mountains near Gumbum monastery, May 20, 1890—Gr.-Grzh.; Humboldt mountain range near Kuku-Usu area, 2750–3650 m, near brook on sand, May 26, same site, nor. slope, Ulan-Bulak area, montane gorge, 3650 m, June 15; Ritter mountain range, 3050–3950 m, placers and meadow alongside brook, June 28; Humboldt mountain range, Ulan-Bulak gorge, alp. belt, 3950 m, June 29; Sharagoldzhin valley (Dankhe river), 3050–3350 m, along arid sandy pebbly bed, July 6–1894, Rob.; Cheibsen-Khit temple, 1901; No. 67a—Lad.; Pabatson' area [Tetung river valley], July 24; Kuku-Nor lake, Uiyu area, Sep. 12—1908, Czet.; Pinfan town, along river bank, July 21, 1909—Czet.; pass through Altyntag along highway from Aksai, 3460 m, Aug. 2, 1958—Petr.; Kuku-Nor lake, east. bank, 3210 m, sandy-pebbly coastal strip, Aug. 5, 1959—Petr.; "Koko-nor, in der Nahe des Sees Ganhadaliannor, um 3150 m; auf dem Gebirge Jahemari, um 4000 m, 1930"-Hao l.c.), *Amdo* (along Baga-Gorgi river on sandy-pebbly bed, May 24; in clayey desert along Huang He river, 2450–2750 m, May 30—1880, Przew.; "sumpfiges Tal, July 6, 1904, Filchner"—Diels l.c.).

IIIB. Tibet: *Chang Tang* (nor. slope of Russky mountain range, Kyutel'-Dar'ya river, 3650 m, along brook on gravel, May 15; south. slope of Russky mountain range, 3350–3650 m, along brooks on gravel, July 2—1890, Rob.), Norra Tibet, Lager XIII, Kaltha-alaghan, 4652 m, July 24, 1900—S. Hedin, fl., syntypus *O. hedinii* [S, C]; "nordfluss des Kizil-dawan, felsiges Tal des Kurabsu, 2950 m, June 18; gleiche Lokalitat, 3000 m, June 20; Ullug Kul, 4950 m, June 26; Baba Hatun, Keria Daria, 5200 m, June 29; oberer Keria Daria, grasreiche mit Lavablochen, uberstreute Talebene, 5170 m, July 7—1907, Zugmayer"—Hand.-Mazz. l.c.; Kun'-Lun', nor. slope in Keriya river basin, 17 km south and beyond Polur, 3000 m, May 12, 1959—Yun. and I-f. Yuan'), *Weitzan* (Alyk-Norin-Gol river valley, Aug. 11, 1884—Przew.; Burkhan-Budda mountain range, nor. slope, Nomokhun gorge, May 20, 1890; Alyk-Norin-Gol river valley, June 8, 1901—Lad.), *South.* ?

IIIC. Pamir (Ostra Pamir, Mustagh-ata, Tergen-bulak, gletschertunga, 4374 m, Aug. 14, 1894—S. Hedin fr., syntypus *O. hedinii* [S, C]; Kun'-Lun', Tynnen-Daban pass, 4000–4200 m, June 26, 1942—Serp.).

General distribution: East. Pamir (Shotput area); China (North-West: south. Gansu).

Note. 1. A comparison of type specimens of *O. falcata* Bge. and *O. hedinii* Ulbr. did not reveal any differences. On the label of the herbarium specimens of *O. falcata* collected by G.N. Potanin along Saksa river in Mong. Altay on July 7, 1877, A. Bunge himself noted that this species occupied an intermediate position between sections Polyadena and Gloeocephala because of the predominance of leaves with paired, not whorled, leaflets. The presence of imparipinnate leaves evidently led to the confusion of E. Ulbrecht (l.c.) who assigned his species to section Gloeocephala.

2. Pods measure up to 5 cm long and 8 mm broad.

140. O. microphylla (Pall.) DC. Astrag. (1802) 83; Turcz. Fl. baic.-dahur. 1 (1845) 311; Bge. Sp. Oxytr. (1874) 154; Sap. Mong. Altai (1911) 364; B. Schischk. in Fl. SSSR, 13 (1948) 216; Grub. Konsp. fl. MNR (1955) 191; Fl. Intramong. 3 (1977) 230; Ulzij. in Grub. Opred. sosud. rast. Mong. [Key to Vascular Plants of Mongolia] (1982) 167. —*Phaca microphylla* Pall. Reise, 3 (1776) 744. —**Ic.:** Pall. Reise, 3, tab. 10, fig. 1; Fl. Intramong. 3, tab. 117.

Described from Baikal region. Type in London (BM). Isotype in St.-Petersburg (LE).

In montane arid and desert steppes, solonetzic basins, solonetzic sand and sandy-pebbly banks of lakes and rivers, in chee grass thickets.

IA. Mongolia: *Mong. Alt.* (Tsitsirin-Gol, on arid pebble bed, July 10; lower Dzhirgalantu, July 16; [Adzhi-Bogdo], upper part of Dzusylan gorge, June 29—1877, Pot.; between Dayan-Gol and Akkorum lakes, June 29; Borbolgasun river, July 2—1903, Gr.-Grzh.; arid meadows along Tsagan-Gol, steppe slopes, June 21, 1906—Sap.; east. slopes of Tonkhilnur depression, montane desert steppe, Aug. 13, 1945; west. bank of Tonkhil-Nur along road crossing through cape of a mud cone of small lake, July 16; 12 km nor. of Tamchi somon, feather grass–wheat grass steppe on finely hummocky area, July 16, 1947—Yun.; Bayan-Undur-Nur mountain range near pass 16 km from Bayan-Undur somon along road to Bayan-Tsagan somon, montane steppe, Aug. 26, 1948—Grub.; Bayan-Nuru, nor. slope 1 km before pass, 2250 m, rocky steppe slope, June 22; Tonkhil somon, Dzuilin-Gol valley 5 km from centre of somon along road to Tsastu-Bogdo-Ul, solonchak lowland with gravelly soil, June 23; west. rim of Tsetseg-Nur basin, Temetiin-Khukh-Ula, south-west. slope along road to Must somon, 2150–2200 m, desert-steppe slopes, June 6; upper Kobdo river, Khoton-Gol near exit from Dayan-Nur lake, in solonetzic basin, July 10; interfluvine region of Kobdo river and Sagsai-Gol, Dabatuin-Daba pass on Tsengel somon—Ulan-Khusu road, 2250 m, montane steppe, July 12—1971, Grub., Ulzij. et al.), *East. Gobi* (in salsis Mongoliae chinensis, 1831—I. Kuznetsov; Gurban-Saikhan somon, 12–15 km south of Tabchin-Chzhis, depression between mud cones, feather grass steppe on solonetzic soil, June 23, 1949—Yun.).

General distribution: *East. Sib.* (Daur.).

Note. In Claves pl. Xinjiang. 3 (1985) 96, this species was reported from Barlyk and Maili mountain ranges in Chinese Junggar although the author (Yang Chang-You) did not study at all the Sinkiang specimens and this report is obviously erroneous.

O. muricata (Pall.) DC. Astrag. (1802) 86; Turcz. Fl. Baic.-dahur. 1 (1945) 311; Bge. Sp. Oxytr. (1874) 153; Kryl. Fl. Zap. Sib. [Flora of West Siberia] 7 (1933) 1764; B. Schischk. in Fl. SSSR, 13 (1948) 215; Grub. Konsp. fl. MNR (1955) 191; Ulzij. in Grub. Opred. sosud. rast. Mong. [Key to Vascular Plants of Mongolia] (1982) 167. —*Phaca muricata* Pall. Reise, 3 (1776) 318 adn 746.

Described from East. Siberia. Type in London (BM).

Found in Nor. Mongolia [Fore Hubs., Mong.-Daur.) and eastern regions of Altay but not found within Cen. Asia. Erroneously reported (Grub. l.c., Ulzij. l.c.) from East. Mong. and Gobi-Alt.

141. **O. trichophysa** Bge. Sp. Oxytr. (1874) 158; Saposhn. Mong. Altai (1911) 364; Kryl. Fl. Zap. Sib. 7 (1933) 1765; B. Schischk. in Fl. SSSR, 13 (1948) 214; Grub. Konsp. fl. MNR (1955) 194; Fl. Kazakhst. 5 (1961) 407; Ulzij. in Grub. Opred. sosud. rast. Mong. [Key to Vascular Plants of Mongolia] (1982) 167; Claves pl. Xinjiang. 3 (1985) 96.

Described from Mong. Altay. Type in St.-Petersburg (LE). Plate IV. fig. 3,

Along steppe and desert rocky and rubbly slopes of mountains and river valleys, in ravines, on arid pebble beds of rivers and gorges.

IA. Mongolia: *Khobd.* (Iter ad Kobdo, Fluss Boroburgassi, June 17, 1870—Kalning, typus!; mountains on left bank of Kharkhira river, July 11; mountains between left bank of Kharkhira river and Tszusylan area, July 11; Kharkhira river valley, July 20; Kharkhira river top, 2450 m, July 22—1879, Pot.; Kashgachsk steppe up to upper Kobdo river, spring and summer 1897, S. Demidova), *Mong. Alt.* (Tsitsirin-Gol, in a subsidiary valley on rocky soil, July 9; same site, on arid pebble bed, July 10; Yamadzhin mountains, on rocky soil, July 11—1877, Pot.; Kak-Kul' lake between Tsagan-Gol and Kobdo, arid rubbly ravines, June 22, 1906—Saposhn.; "Oigur, Chary-Khargai, Saksai, Bzau-Kul', Terekty, Shiverin-Gol, alpine barren steppe"—Saposhn. l.c.; mountains between Kengurlen and Tarkhite rivers, July 7; at Malyi Ulan-Daban pass, July 18—1898, Klem.; Khara-Adzarga mountains, Sakhir-Sala river valley, south. rubbly slope, Aug. 23, 1930—Pob.; left bank slope in Indertiin-Gol valley near summer camp in Bulgan somon, rocky steppe, Aug. 25, 1947—Yun.; Mong. Altay 40 km nor.-west of Dzakhoi oasis, 2300 m, montane steppe, Aug. 16, 1973—Isach. and Rachk.), *Depr. Lakes* (Dzun-Dzhirgalantu mountain range, south-west. slope, 1850—2800 m, Ulyastyn-Gola gorge, June 28, 1971—Grub., Ulzij. et al.), *Val. Lakes* (rather close to Baidarik river, before Dzag-Baidarik [somon], Aug. 6, 1930–Pob.), *Gobi-Alt.* (Baga-Bogdo, walls and canyon bottoms at 1850 m, No. 205, 1925—Chaney; Ikhe-Bogdo mountain range, offshoots, upper belt, June 29, 1926—E. Kozlova; Arts-Bogdo, nor. macroslope, 2000 m, on flank of ravine, July 4, 1973—Isach. and Rachk.).

IIA. Junggar: *Tien Shan* (upper Taldy, 2150 m, May 15; same site, 2150–2450 m, May 16; same site, 2150 m, May 22; Irenkhabirga, Taldy, 2450 m, May 24; south. sources of Taldy, 2450–3050 m, May 25; Dzhin, 1500 m, June 6; Irenkhabirga, Tsaganusu, 1850–2450 m, June 10—1879, A. Reg.), *Jung. Gobi* (east.: Baitak-Bogdo mountain range, nor. macroslope, 2600 m, montane steppe, Aug. 8, 1977—Volk. and Rachk.).

General distribution: *West. Sib.* (Altay, upper Chui river).

Section 9. Gloeocephala Bge.

142. **O. fragilifolia** Ulzij. in Bot. zh. 64, 9 (1979) 1234; Ulzij. in Grub. Opred. sosud. rast. Mong. [Key to Vascular Plants of Mongolia] (1982) 164. —Ic.: Grub. l.c. Plate 87, fig. 398.

Described from Mong. Altay. Type in St.-Petersburg (LE).

On rocky and stony slopes, rubbly talus and rock screes in alpine and montane steppe belts, 2000–3200 m alt.

70   IA. Mongolia: *Mong. Alt.* (Khasagtu-Khairkhan mountain range, upper gorge of Khunkeriin-Ama under summit of Tsagan-Irmet, 2980 m, on southern slope, on rubbly placer, Aug. 23, 1972—Grub., Ulzij., Tsetsegma, typus!; Burkhan-Budai-Ula, nor. macroslope, alpine region, 3000 m, June 19; same site, 2000 m, on slope of arid gully, June 19—1973, Banzragch and Munkhbayar; Adzhi-Bogdo, south macroslope, Ikhe-Gol river, 3100–3200 m, north-exposed flank of gorge, talus slope, Aug. 22, 1979, Grub., Dariima et al.), *Gobi Alt.* (Ikhe-Bogdo, south. macroslope, Bityuten-Ama creek valley, right offshoot, subalpine montane steppe, Sept. 12, 1943; same site, upper creek valley of Narin-Khurimt, steppe belt in upper part of nor. slope, June 28, 1945—Yun.; Baga-Bogdo, Bambuguriin-Ama gorge, under pass, on steep rocky slope exposed northward, Aug. 21, 1967—Ulzij., Avgangongor).

General distribution: endemic.

Section 10. Leucopodia Bge.

143. **O. squamulosa** DC. Astrag. (1802) 79; Turcz. Fl. baic.-dahur. 1 (1845) 313; Bge. Sp. Oxytr. (1874) 130; Saposhn. Mong. Altai (1911) 363; Kryl. Fl. Zap. Sib. [Flora of West. Siberia] 7 (1933) 1756; Peter-Stib. in Acta Hotri Gotob. 12 (1937) 82; B. Fedtsch. and Vassilcz. in Fl. SSSR, 13 (1948) 213; Grub. Konsp. fl. MNR (1955) 193; Hanelt and Davazamc in Feddes Repert. 70, 1–3 (1965) 44; Fl. Intramong. 3 (1977) 220; Ulzij. in Grub. Opred. sosud. rast. Mong. [Key to Vascular Plants of Mongolia] (1982) 164. —*O. leucopodia* Ledeb. Fl. Alt. 3 (1837) 279; Opred. rast. Tuv. ASSR [Key to Plants of Tuva Autonomous Soviet Socialist Republic] (1984) 152. —Ic.: DC. l.c. tab. 3; Ledeb. Ic. pl. fl. ross. 3, tab. 282; Fl. Intramong. 3, tab. 112, fig. 7–12.

Described from Dauria (?). Type in Paris (P). Plate IV, fig. 5.

On rubbly desert-steppe slopes of mountains and knolls, along solonetzic floors of lowlands, sandy and pebbly beds of gorges, on pebble bed terraces of rivers.

IA. Mongolia: *Mong. Alt.* (Kak-Kul' lake, standing moraine, July 17, 1909—Saposhn.; "Tsagan-Gol"-Sap. l.c.; Buyantu river basin, west. offshoot of Bugu-Ula, 2250 m, montane desert steppe, July 2, 1971—Grub., Ulzij., Darrima), **Cen. Khalkha** (Choiren-Ula, 1940—S. Damdin; Gal-Shara somon, Gashyunyi-Gol area, sand bed of gorge, May 20, 1940—Yun.), *East. Mong.* (nor.-west. fringe of Zodol-Khan-Ula, steppe on basalts, May 14; 15 km nor.-nor.-west of Baishintu-Sume, floor of steppe valley, May 18—1944, Yun.), *Depr. Lakes* (south. fringe of Ubsu-Nur lake basin 6 km east of Turgen-Gol along road to Malchin somon, taro-white wormwood desert, July 19, 1971—Grub., Ulzij., Dariima), *Gobi Alt.* (Ikhe-Bogdo mountain range, nor. slope, midportion of Bityuten-Ama creek valley, juniper groves on steep slopes, Sep. 12, 1943—Yun.), *East. Gobi* (along road to Tsatkholun well from Bolergen well, on dry bed, May 31, 1909—Napalkov; west of Shara-Murun, on dry plateau, No. 51a, 1925—Chaney; Ongiin-Gol near Khushu-Khit, rubble terrace above floodplain, July 18, 1926—Bulle; Delger-Khangai, around granite rocks, on gravel, July 30, 1931—Ik.-Gal.; 17 km nor. of Dzamyn-Ude, Khukh-Tologoi well, along slopes of Khukh-Tologoi mountains, Aug. 28; 40 km nor.-east of Argaleul mountains near Ovanso-Khul' well, sandy banks of dry river bed, Sep. 5—1931, Pob.; 36 km south-east of Khara-Airik somon along road to Sain-Shandu, finely hummocky area, feather grass steppe, Aug. 27, 1940—Yun.; Bailinmyao, desert steppe, 1959—Ivan.; "Cecij-ul, Berghang, No. 2063", May 1962—Hanelt and Davazamc l.c.; Shankhai-Nuru mountains near Dashlingiin-Khure ruins, along gorge floor, July 27, 1989—Grub., Gub., Dariima).

General distribution: *West. Sib.* (Altay), East. Sib. (Daur., Sayans), Nor. Mong. (Hangay: Khan-Khukhei mountain range).

Subgenus 3. **PHYSOXYTROPIS** Bge.

144. **O. bungei** Kom. in Feddes Repert. 13 (1914) 229; Pavl. in Byull. Mosk. obshch. ispyt. prir., otd. biol. 38, 1–2 (1929) 95; Grub. Konsp. fl. MNR (1955) 188; Hanelt and Davazamc in Feddes Repert. 70, 1–3 (1965) 44; Ulzij. in Grub. Opred. rast. Mong. [Key to Plants of Mongolia] (1982)

169. —Ic.: Grub. Opred. sosud. rast. Mong. [Key to Vascular Plants of Mongolia] Plate 86, fig. 397.

Described from Mong. Altay. Type in St.-Petersburg (LE). Map 1.

On rocky and rubbly steppe slopes and trails of mountains and knolls, talus and rocks.

71   IA. Mongolia: *Mong. Alt.* (Uzun-Dzyur [Dzuilin-Gol] river, 2 verstas (1 versta = 1.067 km) beyond Da-Guna Khure, on talc-clay-shale cliff, July 28, 1896—Klem.; typus!; Khasagtu-Khairkhan mountain range along Gobi-Altay—Daribi road, rocky slopes, June 20; Bayan-Nuru, nor. slope 1 km before pass, 2250 m, steppe rocky slope, June 22; west. fringe of Tonkhil-Nur basin south of Kholbo-Ulan-Ula, finely hummocky area, rocky steppe slopes, June 25—1971, Grub., Ulzij., Dariima; Adzhi-Bogdo mountain range, south. macro slope, Ikhe-Gol river gorge, 2200 m, montane steppe, Aug. 23, 1979—Grub. and Dariima), *Gobi Alt.* (near Naryn-Bulak spring on south. foothill of Tostu mountain range, Aug. 16, 1886—Pot.; nor. foothill of Ikhe-Bogdo mountain range, rocky subdesert, May 8, 1926—E. Kozlova; Khalga pass between Barun-Saikhan and Dundu-Saikhan mountains, on rubbly slope, Aug. 3, 1931—Ik.-Gal.; "Dund-Sajchan, Bognin-Chjar, Berghange, No. 2075", June 1962—Hanelt and Davazamc l.c.; low montane massif 60 km east of Bayan-Tsagan somon, steppe slope, Aug. 4; Adagin-Khabtsagaitin-Nuru mountain range 10 km nor. of Shine-Zhinst somon, petrophyte-feather grass steppe on nor. slope of mud cone, Aug. 6—1973, Isach. and Rachk.), *East. Gobi* (Del'ger-Hangay mountains, along slope, on rubbly placers, Aug. 1, 1931—Ik.-Gal.).

General distribution: *Nor. Mong.* (Hang., main mountain range, south. slope).

Subgenus 4. PTILOXYTROPIS Bge.

145. O. bella B. Fedtsch. ex O. Fedtsch. in Acta Horti Petrop. 21, 3 (1903) 303; Vassilcz. and B. Fedtsch. in Fl. SSSR, 13 (1948) 227; Fl. Kirgiz. 7 (1957) 393; Ikonnik. Opred. rast. Pamira [Key to Plants of Pamir] (1963) 173; Fl. Tadzh. 5 (1978) 494; Filim. in Opred. rast. Sr. Azii [Key to Plants of Mid. Asia] 7 (1983) 367; C. Yang in Claves pl. Xinjiang. 3 (1985) 74. —*O. tatarica* auct. non Cambess. ex Bge.; Bge. in Mem. Ac. Sci. St.-Petersb. 7 ser. 22, 1 (1874) 16 quoad pl. kokanica.

Described from East. Pamir (Kara-Kul' lake). Type in St.-Petersburg (LE).

On arid rocky and rubbly slopes of mountains, sandy-pebbly terraces of rivers in middle and upper belts of mountains, up to 5000 m.

IIIC. Pamir ("west. fringe of Tashkurgan"—C. Yang. l.c.).
General distribution: East. Pam.; Mid. Asia (Pam.-Alay).

146. O. sacciformis H.C. Fu in Acta phytotax. sin. 20, 3 (1982) 311. —*O. lavrenkoi* Ulzij. in Byull. Mosk. obshch. ispyt. prir., otd. biol. 92, 5 (1987) 114. —Ic.: H.C. Fu l.c. 312.

Described from Mongolia (East. Gobi). Type in Huhhot (NMU) [Khukh-Khoto (NMU)].

On rocky and rubbly desert slopes of hills and knolls.

IA. **Mongolia:** *East. Gobi* (Ulanqab Meng. Darhan Muminggan Lianhegi, Bailingmiao, in collibus, Sep. 5, 1979, S.Y. Zhao, No. 35—typus! NMU; 200 km south of Sain-Shanda town, along nor. slope of Khutag-Ula hills, 1110–1400 m, No. 5830, June 19, 1980—Gub., typus *O. lavrenkoi!*; same site, Khutag-Ula, on slope, 1300 m, July 10, 1982—Gub.).

General distribution: endemic.

Note. Similar to *O. trichosphaera* Freyn. in habit but peduncles and rachis rigescent, long-persistent, and plant spiny. The author of this species (H.C. Fu) placed it in subgenus *Physoxytropics* Bge. and established a new section *Mongolia* H.C. Fu l.c. but the species of this subgenus are low, densely caespitose herbs with subsessile flowers while this species with long peduncles and in all other respects belongs to subgenus *Ptiloxytropis* Bge.

147. **O. trichosphaera** Freyn in Bull. Herb. Boiss. 2 ser. 6, 3 (1906) 193; Vassilcz. and B. Fedtsch. in Fl. SSSR, 13 (1948) 226; Ikonnik. Opred. rast. Pamira [Key to Plants of Pamir] (1963) 173; Fl. Tadzh. 5 (1978) 494; Filim. in Opred. rast. Sr. Azii [Key to Plants of Mid. Asia] 7 (1983) 367; Claves pl. Xinjiang. 3 (1985) 74. —**Ic.:** Fl. Tadzh. 5, Plate 66, fig. 13–18.

Described from East. Pamir (Yashil'-Kul' lake) Type in Geneva [Geneva] (G). Isotype in St.-Petersburg (LE).

72 On rocky and rubbly arid slopes in middle and upper belts of mountains.

IIIC. **Pamir** ("west. extremity of Tashkurgan"—Claves pl. Xinjiang. l.c.).

General distribution: East. Pam.; Mid. Asia (Pam.-Alay).

Note. *O. trichocalycina* Bge., closely related to this species, was reported from Sinkiang (Claves pl. Xinjiang. 3: 74) but it is evidently incorrect: this is endemic in West. Tien Shan and west. Pamiro-Alay.

Subgenus 5. **TRITICARIA** Vass.

148. **O. hirta** Bge. Enum. pl. China bor. (1832) 17, No. 101; id. Sp. Oxytr. (1874) 120; Forbes and Hemsley, Index Fl. Sin. 1 (1887) 167; Peter-Stib. in Acta Horti Gotob. 12 (1937) 78; Boriss. in Fl. Zabaik. [Flora of Transbaikalia] 6 (1954) 598; Fl. Intramong. 3 (1977) 223. —*O. komarovii* Vass. in Not. syst. (Leningrad) 20 (1960) 249; Vassilcz. in Fl. SSSR, 13 (1948) 228; Grub. Konsp. fl. MNR (1955) 190; Ulzij. in Grub. Opred. rast. Mong. [Key to Plants of Mongolia] (1982) 169. —**Ic.:** Fl. Intramong. 3, tab. 114, fig. 1–7; Grub. Opred. rast. Mong. [Key to Plants of Mongolia], Plate 89, fig. 407 (sub nom. *O. komarovii*).

Described from North China (Beijing environs). Type in Paris (P). Isotype in St.-Petersburg (LE). Plate IV, fig. 1.

On steppe slopes of mountains and knolls, in meadow and sandy steppes.

IA. **Mongolia:** *East. Mong.* (in locis subarenosis Mongoliae chinensis, 1831—I. Kusnezov, Bge.; Khaligakha area, June 23; Ikhyr lake, July 6—1899, Pot. and Sold.;

Shilin-Khoto, feather grass-wild rye steppe, 1959—Ivan.; Barun-Sul hill, herbage-common cattail steppe with apricot, July 1, 1971—Dashnyam, Isach. et al.; Derkhin-Tsagan-Obo mountain 35 km south-east of Khamar-Daba settlement, sandy southern slope, steppe, July 26, 1971—Karam. and Safron.).

General distribution: *East. Sib.* (Daur.), Nor. Mong. (Fore Hing.), China (Dunbei, North-West).

Note. *O. komarovii* Vass. is no different from *O. hirta* Bge.; cusp length of keel varies greatly, usually shorter than 2 mm, even among Beijing plants and bracts of Daurian and Manchurian plants have a single nerve, as in Beijing plants. These, therefore, cannot be differentiated on the basis of characteristics indicated by I.T. Vassilczenko into 2 species.

Subgenus 6. TRAGANTHOXYTROPIS Vass.[1]

Section 1. Lycotriche Bge.

149. O. aciphylla Ledeb. Fl. alt. 3 (1831) 279; Bge. Sp. Oxytr. (1874) 134; Pavl. in Byull. mosk. obshch. ispyt prir., otd. biol. 38, 1–2 (1929) 95; Kryl. Fl. Zap. Sib. [Flora of West. Siberia] 7 (1933) 1759; Peter-Stib. in Acta Horti Gotob. 12 (1937) 82; Vassilcz. and B. Fedtsch. in Fl. SSSR, 13 (1948) 225; Grub. Konsp. fl. MNR (1955) 187; Fl. Kazakhst. 5 (1961) 410; Hanelt and Davazamc in Feddes Repert. 70, 1–3 (1965) 44; Fl. Intramong. 3 (1977) 218; Ulzij. in Issled. fl. i rast. MNR [Study of Flora and Vegetation of Mongolian People's Republic] 1 (1979) 107; id. in Grub. Opred. rast. Mong. [Key to Plants of Mongolia] (1982) 163; Filim. in Opred. rast. Sr. Azii [Key to Plants of Mid. Asia] 7 (1983) 367; Opred. rast. Tuv. ASSR [Key to Plants of Tuva Autonomous Soviet Socialist Republic] (1984) 150; Claves pl. Xinjiang. 3 (1985) 76. —Ic.: Ledeb. Ic. pl. fl. ross. 3, tab. 281; Fl.
73 SSSR, 13, Plate 9, fig. 1–3; Fl. Kazakhst. 5, Plate 50, fig. 1; Fl. Intramong. 3, tab. 109, fig. 1–6.

Described from Altay (Irtysh river). Type in St.-Petersburg (LE).

On plain and hummocky sand, sandy-rubbly trails and slopes of mountains, sandy-pebbly floors of gorges; characteristic of fruticulose desert and desert-steppe associations.

IA. Mongolia: *Khobd., Mong. Alt., Depr. Lakes, Val. Lakes, Gobi Alt., East. Gobi* (Mongolia chinensis [Kalgan road between Barrun-Mingan and Shilin Khuduk] 1831—Bunge, Rozov; between Mingan and Sain-Usu, May 28 [June 9] 1841—Tatarinow; nor. Inshan', on sand mixed with clay 4 ft in diam. and 2 ft high, from mid-May to

---

[1]Unlike subgenera established for *Oxytropis* DC. A. Bunge, this subgenus, in our opinion, is artificial based on biomorphic and not accepted taxonomic characteristics like structure and ratio of calyx to pod, leaf structure (imparipinnate or whorled) and bracts, degree of development of stem and peduncle, type of inflorescence etc. Based on these characteristics, the species covered under subgenus Traganthoxytropis Vass. are entirely heterogeneous.

mid-June, 1872—Przew.; Khara-Obo well, Sep. 18 [30] 1880—Przew.; Galbyn-Gobi, Ulan-Tatar area, sandy-rocky soil, May 10, 1909—Napalkov; Baga-Ude, ravine in Khara-Ula mountains, along floor of ravine on gravel, Aug. 6, 1926—Lis.; Kalgan road, from Dolo-Chelut ridges to Ude well, subdesert, Aug. 24, 1931—Pob.; Ail-Bayan somon, 25 km from Shine-Usu, along road to Ulgii somon, turf-covered sand in solonchak valley, Sep. 19; Khan-Bogdo somon, nor. fringe of fine-hummocky Galba, on rocky slope among rocks, Sep. 28—1940; Ikhe-Dzhargalan somon, 15 km south-south-west of somon along road to Under-Shili, feather grass desert steppe, June 1, 1941—Yun.; Bail'nmyao town, desert feather grass steppe, 1959—Ivan.; 1 km west of No. 43 railway siding, Reaumuria-Saltwort association on solonetzes, July 20, 1971—Isach. and Rachk.), Alash. Gobi (Alashan, Oct. 1, 1871; Muni-Ula, left bank of Huang He and nor. Alashan hills, from mid-April to May—1872; Alashan, June 20—July 10, 1873—Przew.; Tengeri sand, Deidzikho area, Sep. 23, 1901—Lad.; Dyn'yuan'in, on clayey-rocky soil and on talus, April 4; Tengeri sand, Tarbagai area, among dunes, July 7—1908; Dyn'yuan'in, south. slope of clayey mountains, April 10; Alashan, Shara-Burdu area, May 5—1909, Czet.; Nomogon somon, Onchi-Khyar area 40–45 km south-south-west of Bulgan khural, on slope of intensely rocky knoll, April 21, 1941; Bordzon-Gobi area, nor. trail and surrounding finely hummocky region of Khalzan-Ula mountains, shrubby desert along gullies, June 18, 1949; upper part of south. trail of Khurkhe mountain range, along small banks among saltwort deserts, Sep. 8, 1950—Yun.; Chzhunvei, 90 km south-west of town, sand-covered dry bed, June 30, 1957; Bayan-Khoto, 12 km south-west of Taodaokhe, near Syrkhe well, ridge of sand-blown host rocks, June 15, 1958—Petr.; Yuan'tsyuan' along road to Chzhunvei, desert sand, on knolls, June 30, 1957—Kabanov), Ordos (Ordos, Aug. 1871—Przew.; Khuamachen town, Sep. 28; Khoikhoi-fantsza village [west of Linchzhou town], on sand, Oct. 1; Shigui village on southern fringe of Ordos, Oct. 10—1884, Pot.; along road from Sykuz village through Ningan'pu to Dzindzipu, April 20–28, 1909—Napalkov; 15 km from San'shingun town, high sandy-pebbly terrace on right bank of Huang He, June 7, 1958—Petr.).

IB. Qaidam: Plain (Syrtyn 25 km south of Khuakhatsza settlement, nor. submontane sandy-pebbly plain of Mushketov mountain range, Oct. 7; 14 km nor. of Tsagan-Us, half-covered with sand, Oct. 14—1959, Petr.), montane (Nor. Qaidam, Mandzhu-Bulak, in high sandy-clayey valleys, Sep. 4, 1879—Przew.; Serik area, 3350 m, on sand, June 5; Sharagol'dzhin river, Baga-Bulak area, 3200 m, on sand, June 15—1894; Sarlyk-Ula mountains, Dzukhyn-Gol river, 2950 m, on sand, May 8, 1895—Rob.).

IIA. Junggar: Jung. Gobi (east.: on sheer bank of Turkul' lake, June 15, 1877—Pot.; Barkul' lake region, along edge of solonchak, No. 2271, Sep. 28, 1957—Kuan; Khubchigin-Nuru mountain range west of Adzhi-Bogdo, along slopes of knolls, Aug. 3, 1947—Yun.; Bodonchin-Gola valley, Khara-Togo-Khuduk well, along nor. slopes of mounds, June 30, 1973—Golubk. and Tsogt).

IIIA. Qinghai: Nanshan (Gomi on upper Huang He, 2750 m, on high pebble bed bank, May 31; near Kuku-Nor lake, 4000 m, July 9—1880, Przew.; Humboldt mountain range, environs of Kuku-Usu area, June 6, 1894—Rob.; sandy desert surrounding northeastern shores of Kokonor, 3200 m, No. 13382, Sep. 1925—Rock), Amdo (Ar'ku village at confluence of Karuch river into Yellow river, on riverine pebble bed, May 5, 1895—Pot.).

General distribution: West. Sib. (Altay), China (Altay, North—Pohuashan hills).

150. O. kossinskyi B. Fedtsch. et Basil. in Izv. Bot. sada AN SSSR, 26 (1927) 117; Pavl. in Byull. mosk. obshch. ispyt. prir., otd. biol. 38, 1–2 (1929) 95; Grub. Konsp. fl. MNR (1955) 190; Ulzij. in Grub. Opred. rast

Mong. [Key to Plants of Mongolia] (1982) 163. —**Ic**.: Grub. Opred. rast. Mong. [Key to Plants of Mongolia] Plate 87, fig. 401.

Described from Mongolia (Cen. Khalkha). Type in St.-Petersburg (LE).

On steppe rubbly and rocky slopes and rocks on mounds.

IA. Mongolia: *Cen. Khalkha* (road to Urgu from Alashan, along road to Irdynin-Gol area from Bain-Barate area, June 12, 1909—Czet.; Tola river area near crossing, rocky summit and slopes of mountains [on Ulan-Bator-Tsetserleg road], No. 740, July 2, 1924—Pavl. typus!; 15 km nor. of Mandal-Gobi ajmaq centre, among sparse feather grass steppe on rocky places of knolls, June 5, 1949—Yun.; 20 km nor. of Santu somon, mountain slopes, feather grass steppe on highly rubbly soil, July 24, 1951—Kal.; 50 km nor.-east of Mandal-Gobi, Summit of knoll, petrophyte-forb association, July 5, 1970—Banzragch, Karam. et al.; 30 km nor.-east of Arbai-Khere along road [to Ulan-Bator], fine-grass snakeweed-feather grass steppe, June 11, 1971—Grub., Ulzij., Dariima; Khar-Khuzhirtyn-Daba 13 km from camp [Undzhul], along gully, July 6, 1974—Golubkova, Tsogt et al.), *Val. Lakes* (on mountain on right bank of Tuin-Gol, July 9, 1893—Klem.; interfluve of Teliin-Gol and Tatsyn-Gol 25 km west-south-west of Khairkhan-Dulan somon along south. Hangay road, wheat grass-feather grass steppe, June 27, 1941—Yun.; 3 km nor.-east of Barun-Bayan-Ulan somon, Ologoi-Nuryn-Khongor mountain range, rocky fine-hummocky area along fringe of Buridu-Nur lakelet, July 3, 1952—Davazamc; 40 km from Khairkhan-Dulan [along road], 1820 m, shrubbly-petrophyte-forb steppe, June 24, 1972—Banzragch et al.), *East. Gobi* (Dalan-Dzadagad town, solonchak meadow along brook, June–July, 1939—D. Surmazhav).

General distribution: West. Sib. (Altay: Chui steppe).

Note. A poorly differentiated race of *O. aciphylla* Ledeb. confined to steppe petrophyte associations.

Section 2. Hystrix Bge.

151. **O. hystrix** Schrenk in Bull.Ac. Sci. St.-Petersb. 10 (1842) 254; Bge. Sp. Oxytr. (1874) 132; Kryl. Fl. Zap. Sib. 7 (1933) 1758; Vassilcz. and B. Fedtsch. in Fl. SSSR, 13 (1948) 222; Fl. Kazakhst. 5 (1961) 408; Filim. in Opred. rast. Sr. Azii [Key to Plants of Mid. Asia] 7 (1983) 366; Claves pl. Xinjiang. 3 (1985) 76. —*O. spinifer* Vass. in Not. syst. (Leningrad) 20 (1960) 249; Vass. in Fl. URSS, 13 (1948) 222, descr. ross.; Fl. Kazakhst. 5 (1961) 408; Filim. in Opred. rast. Sr. Azii [Key to Plants of Mid. Asia] 7 (1983) 366; Claves pl. Xinjiang, 3 (1985) 77.

Described from East. Kazakhstan (Tarb.). Type in St.-Petersburg (LE). Plate I, fig. 2.

On rocky slopes and rocks in steppe belt of mountains.

IIA. Junggar: *Tarb.* ("Dachen"), *Jung. Ala Tau* ("Toli", "Ven'tsyuan'"—Claves pl. Xinjiang. l.c.).

General distribution: Jung.-Tarb.

Note. The length and thickness of spines vary greatly depending on habitat conditions as could be noticed even among authentic specimens of this species. Environmental conditions of vegetation are intensely reflected on the sizes of flowers

74

94

and fruits as well as the development of spines. *O. spinifer* Vass. described from Jung. Ala Tau, differing only in size of spine, flowers and fruits cannot therefore be described as a distinct species; such forms are also found in Tarbagatai region (our own collections!). A similar phenomenon is also noticed among related *O. tragacanthoides* Fisch.

152. **O. tragacanthoides** Fisch. in DC. Prodr. 2 (1825) 280; Ledeb. Fl. alt. 3 (1931) 278; Bge. Sp. Oxytr. (1874) 131; Danguy in Bull. Mus nat. hist. natur. 17, 6 (1911) 270; Pavl. in Byull. mosk. obshch. ispyt prir., otd. biol. 38, 1–2 (1929) 95; Gr.-Grzh. Zap. Mong. [West. Mongolia] 3, 2 (1930) 817; Kryl. Fl. Zap. Sib. [Flora of West. Siberia] 7 (1933) 1757; Peter-Stib. in Acta Horti Gotoburg. 12 (1937) 82; Vassilcz. and B. Fedtsch,. in Fl. SSSR, 13 (1948) 223; Grub. Konsp. fl. MNR (1955) 193; Fl. Kazakhst. 5 (1961) 409; Hanelt and Davazamc in Feddes Repert. 70, 1–3 (1965) 44; Ulzij. in Issled. fl. i rast. MNR [Study of Flora and Vegetation of Mongolian People's Republic] 1 (1979) 108; id. in Grub. Opred. rast. Mong. [Key to Plants of Mongolia] (1982) 164; Filim. in Opred. rast. Sr. Azii [Key to Plants of Mid. Asia] 7 (1983) 366; Opred. rast. Tuv. ASSR [Key to Plants of Tuva Autonomous Soviet Socialist Republic] (1984) 150; Claves pl. Xinjiang. 3 (1985) 76. —*O. paratragacanthoides* Vass. in Novit. syst. pl. vasc. 6 (1969) 153. —**Ic.**: Ledeb. Ic. pl. fl. ross. 3, tab. 270; Fl. Kazakhst. 5, Plate 50, fig. 2; Grub. Opred. rast. Mong. [Key to Plants of Mongolia] Plate 86, fig. 395.

75   Described from Altay. Type in Geneve (Geneva) (G).

On rocks, rocky and stony slopes, talus, along pebble beds of rivers, in montane steppe belt and on rather thin sand in desert steppes.

IA. Mongolia: *Khobd., Mong. Alt., Cen. Khalkha* (easternmost find: Del'ger-Khan massif, 104–105° E. long., granite outliers, Sep. 15, 1925—Gus.), *Depr. Lakes, Val. Lakes, Gobi Alt.* (easternmost find: Dzun-Saikhan mountain range), *West. Gobi* (Atas-Bogdo mountain range, upper belt of desert steppe, Aug. 12, 1943—Yun.; 35 km west of barrier on Segs-Tsagan-Bogdo mountain range, nor. slopes of low granite hills, Aug. 20, 1973—Isach. and Rachk.).

IC. Qaidam: *montane* (Nor. Qaidam, Khabirga spring, 3350–3650 m, on arid bed and mountain slopes, June 3, 1895—Rob.).

IIA. Junggar: *Tarb.* ("Saur"—Claves pl. Xinjiang. l.c.), *Jung. Ala Tau* (40–42 km nor.-east of Junggar exit to Karaganda pass on Maili-Barlyk mountain range, along rubbly southern slope in steppe belt, Aug. 14, 1957—Yun. and I-f. Yuan'), *Tien Shan* (Khalga river valley 25 km nor.-west of Balinte settlement along road to Yuldus from Karashar, steppe belt, along rocky slope, Aug. 1, 1958—Yun. and I-f. Yuan'), *Jung. Gobi* (steppe valley nor. of Tien Shan [Karlyktaga], along Sachzhan river, on sandy soil, May 12, 1877—Pot.).

IIIA. Qinghai: *Nanshan* ([Kuku-Usu] 3350–3950 m, on silty slopes, July 16, 1879—Przew., environs of Kuku-Usu area, 2750–3050 m, along loessial mounds, May 15; same site, 3050–3650 m, along loess and loess with pebble, June 6; Baga-Bulak area, June 15—1894, Rob.; "Woukomiao, 2000 m, rochers, May 25, 1908, Vaillant"–Danguy l.c.; Altyntag mountain range 15 km south of Aksai settlement, 2800 m, rocky gorge slopes, Aug. 2; pass through Altyntag mountain range 24 km from Aksai settlement on highway to Qaidam, 3450 m, Aug. 2—1958, Petr.).

IIIB. Tibet: *Weitzan* (Burkhan-Budda mountain range, south. slope toward Alyk-Nor lake, 3850 m, along dry rocky beds, May 30, 1900—Lad.).

General distribution: *Jung.-Tarb.* (Tarb.); West. Sib. (Altay), East. Sib. (Sayans), Nor. Mong. (Fore Hubs., Hang).

Note. Shows considerable variation in height and size of bush, length and thickness of spines, size of leaflets, fruits and flowers depending on habitat conditions.

Section 3. **Acanthos** Ulzij.

153. **O. acanthacea** Jurtz. in Novit. Syst. Pl. Vasc. (1964) 203; Ulzij. in Bot. Zhurn. 64, 9 (1979) 1228; id. in Grub. Opred. rast. Mong. [Key to Plants of Mongolia] (1982) 164; Opred. rast. Tuv. ASSR [Key to Plants of Tuva Autonomous Soviet Socialist Republic] (1984) 150. —Ic.: Grub. l.c., Plate 86, fig. 396.

Described from Altay (Tsagan-Shibetu mountain range). Type in St.-Petersburg (LE).

On rocks, rocky and stony slopes, talus.

IA. Mongolia: *Khobd.* (Tsagan-Shibetu mountain range, east. extremity 1 km north of Ulan-Daba, rocky slope, Aug. 17, 1968—Ulzij., Sanzhid; Turgen' mountain range, Turgen'-Gola gorge 5 km away from estuary, right bank, shore rocks, July 16, 1971—Grub., Ulzij., Dariima), *Mong. Alt.* (Tolbo-Nur somon, Tsast-Ula 3 km north of Bayan-Enger pass [Namarzan], rocky slope, Aug. 12, 1968—Ulzij.).

General distribution: *West. Sib.* (east. Altay).

Plate I.
1—*Oxytropis kansuensis* Bge.; 2—*O. hystrix* Shrenk.; 3—*O. globiflora* Bge.;
4—*O. coerulea* (Pall.) DC.

Plate II.
1—*Oxytropis eriocarpa* Bge.; 2—*O. monophylla* Grub.; 3—*O. deflexa* (Pall.) DC.;
4—*O. strobilacea* Bge.

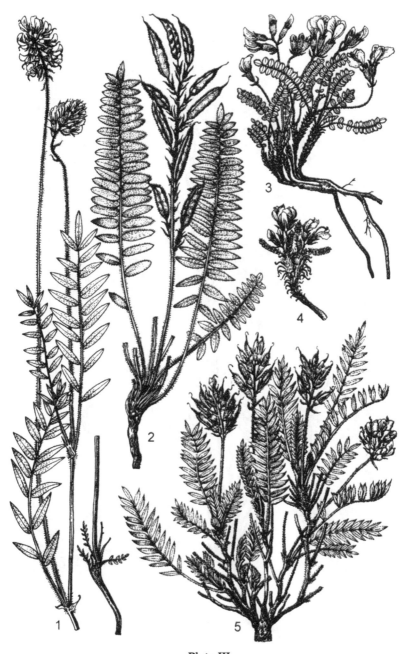

Plate III.
1—*Oxytropis meinshausenii* C.A. Mey.; 2—*O. macrocarpa* Kar. et Kir.;
3—*O. dichroantha* Schrenk; 4—*O. sutaica* Ulzij.; 5—*O. hirsuta* Bge.

Plate IV.
1—*Oxytropis hirta* Bge.; 2—*O. racemosa* Turcz.; 3—*O. trichophysa* Bge.;
4—*O. lasiopoda* Bge.; 5—*O. squamulosa* DC.

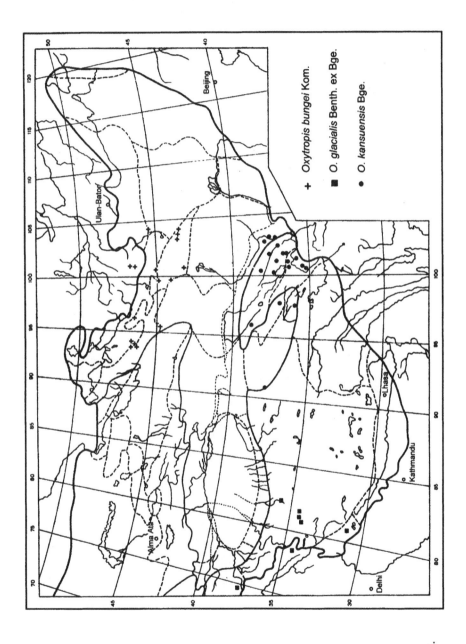

Map 1.

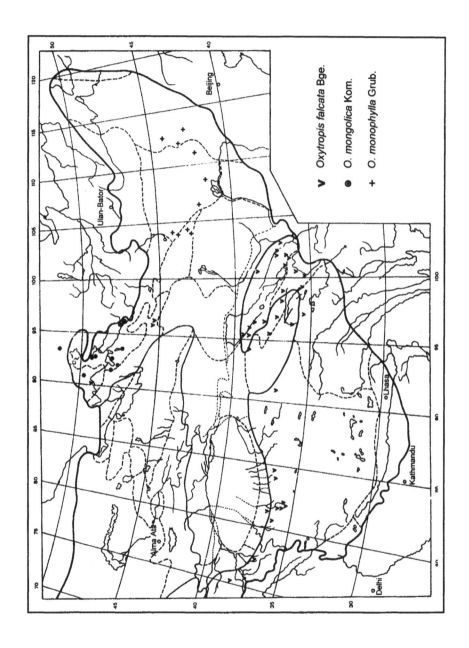

Map 2.

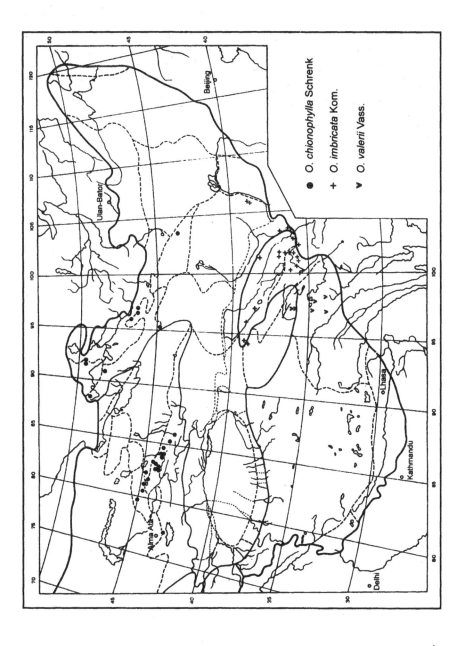

Map 3.

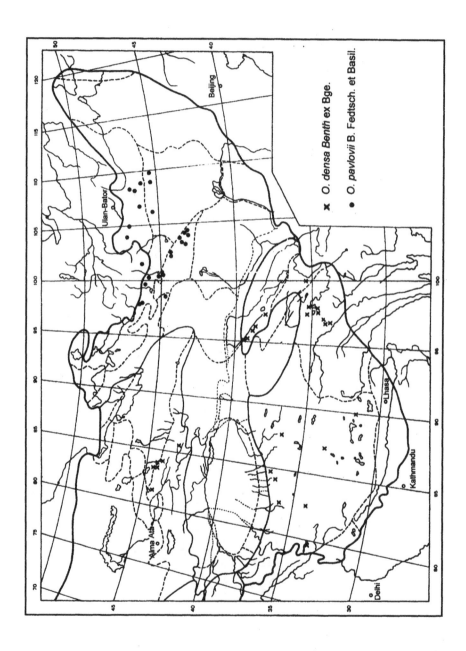

Map 4.

INDEX OF LATIN NAMES OF PLANTS

Acanthos Ulzij., sect. (Oxytropis) 6, 75
Astragalus altaicus Pall. 42
— *ambiguus* Pall. 42
— *ampullatus* Pall. 50
— *argentatus* Pall. 42
— *baicalensis* Pall. 25
— *coeruleus* Pall. 25
— *dahuricus* Pall. 62
— *deflexus* Pall. 33
— *floribundus* Pall. 48
— *glaber* Lam. 34
— *grandiflorus* Pall. 44
— *immersus* Baker ex Aitch. 27
— *leptophyllus* Pall. 53
— *Loczii* Kanitz. var. *scaposa* Kanitz. 27
— *myriophyllus* Pall. 59
— *oxyphyllus* Pall. 61
— *pilosus* L. 39
— *retroflexus* Pall. 33
— *setosus* Pall. 56
— *soongoricus* Pall. 46
— *verticillaris* L. 59
Baicalia(Stell.) Bge., sect. (Oxytropis) 17, 56, 64
Chrysantha Vass., sect. 36
Dolichanthos Gontsch., sect. 39
Eumorpha Bge., sect. 13, 39
Eumorpha (Bge.) Abduss., subgen. 7, 29, 36, 39
Gloecephala Bge., sect. (Oxytropis) 12, 68, 69
Gobicola Bge., sect. 17, 65
Hystrix Bge., sect. 6, 74
Ianthina Bge., sect. 7, 22, 24, 29
Leucopodia Bge., sect. 12, 70

Lycotriche, sect. 6, 72
Mesogaea Bge., sect. 9, 32, 36
Mongolia H.C.Fu, sect. 71
Orobia (Bge.) Aschers. et Graebn., sect. 13, 29, 41
Ortholoma Bge., sect. 12, 36, 47
Oxytropis DC. 6
— acanthacea Jurtz. 6, 75
— aciphylla Ledeb. 6, 72
— acutirostrata Ulbr. 65
— aequipetala Bge. 22
— aigulak Saposhn. 55
— algida Bge. 48
— alpina Bge. 14, 41
— altaica (Pall.) DC. 14, 42
— ambigua (Pall.) DC. 14, 42
— ampullata (Pall.) Pers. 17, 50
— angustifolia Ulbr. 56
— arenaria Jurtz. 61
— argentata (Pall.) Pers. 14, 42
— argyrophylla Ledeb. 42
— assiensis Vass. 17, 51
— *atbaschi* Saposhn. 29
— *avis* Saposhn. 30
— barcultagi Grub. et Vass. 15, 39
— bella B. Fedtsch. ex O. Fedtsch. 20, 71
— bicolor Bge. 19, 56
— biflora P.C. Li 11, 20
— biloba Saposhn. 13, 47
— *bogdoschanica* Jurtz. 63
— brachybotrys Bge. 49
— brachycarpa Vass. 47
— brevipedunculata P.C. Li 7, 24
— *brevirostra* DC. 42
— *Bungeana* Schischk. 43
— Bungei Kom. 20, 70

Oxytropis Burchan-Buddae Grub. et Vass. 17, 51
— cachemiriana Camb. 10, 32
— caespitosula Gontsch. 15, 40
— cana Bge. 10, 32
— chantengriensis Vass. 8, 25
— chiliophylla Royle 20, 66
— chionobia Bge. 18, 57, 62
— chionophylla Schrenk 14, 42
— chorgossica Vass. 9, 32
— *chrysotricha* Franch. 60
— ciliata Turcz. 16, 51
— *coelestis* Abduss. 31
— coerulea (Pall.) DC. 8, 25
— *collina* Turcz. 44
— confusa Bge. 14, 43
— crassiuscula Boriss. 15, 41
— cuspidata Bge. 15, 40
— *Davidii* Franch. 59
— deflexa (Pall.) DC. 10, 33
— densa Benth. ex Bge. 8, 25
— dichroantha Schrenk 13, 48
— *diffusa* Ledeb. 33, 34
— — α *elongata* 34
— — β *pumila* 34
— diversifolia Peter-Stib. 16, 51
— *diversifolia* auct. non Peter-Stib. 54
— *drakeana* Franch. 33
— dschagastaica Grub. et Vass. 15, 40
— dumbedanica Grub. et Vass. 8, 26
— *elegans* Kom. 58
— eriocarpa Bge. 16, 52
— *ervicarpa* Vved. ex Filim. 30
— falcata Bge. 20, 67, 68
— Fetissovii Bge. 18, 57
— filiformis DC. 9, 26
— floribunda (Pall.) DC. 12, 48
— — var. *brachycarpa* Kar. et Kir. 50
— fragilifolia Ulzij. 12, 69
— frigida Kar. et Kir. 13, 43
— fruticulosa Bge. 13, 48
— Geblerii Fisch. ex Bge. 14, 43
— gerzeensis P.C. Li 10, 33
— glabra (Lam.) DC. 9, 33
— — var. *pamirica* B. Fedtsch. 35
— — var. salina (Vass.) Grub. 9

— glacialis Benth. ex Bge. 12, 20
— *glareosa* Vass. 34
— globiflora Bge. 8, 21, 22
— *Goloskokovii* Bait. 34
— Gorbunovii Boriss. 9, 34
— gracillima Bge. 19, 65
— grandiflora (Pall.) DC. 13, 44
— *Grenardi* Franch. 66
— Grum-Grshimailoi Palib. 13, 48
— Gueldenstaedtioides Ulbr. 10, 35, 36
— *hailarensis* Kitag. 61
— *Hedinii* Ulbr. 67, 68
— heterophylla Bge. 19, 57
— hirsuta Bge. 13, 49
— hirsutiuscula Freyn 9, 35
— hirta Bge. 20, 72
— *holansharensis* H.C.Fu 27, 28
— *Holdereri* Ulbr. 67
— humifusa Kar. et Kir. 8, 26, 27
— — var. *grandiflora* Bge. 29
— hystrix Schrenk 6, 74
— imbricata Kom. 8, 27
— *immersa* (Baker ex Aitch.) Bge.
— *inaria* Ledeb. 63
— *incanescens* Freyn 27
— *ingrata* Freyn 66
— *inschanica* H.C.Fu et Cheng f. 53
— intermedia Bge. 16, 52
— *introflexa* Freyn. 56
— *irbis* Saposhn. 56
— Junatovii Sancz. 16, 52
— *Kanitzii* Simps. 27
— kansuensis Bge. 11, 35, 36, 38
— *kashemiriana* auct. non Cambess. 47
— ketmenica Saposhn. 14, 44
— Klementzii Ulzij. 16, 52
— *Komarovii* Vass. 72
— Kossinskyi B. Fedtsch. 6, 73
— Krylovii Schipcz. 7, 28
— kumbelica Grub. et Vass. 9, 28
— *Lacostei* Danguy 21
— Ladyginii Kryl. 9, 28
— lanata (Pall.) DC. 18, 58
— langshanica H.C.Fu. 18, 58
— lanuginosa Kom. 18, 58

Oxytropis lapponica (Wahl.) J.Gay
10, 36, 38
— — var. *xanthantha* Baker 37
— *lapponica* auct. non Gay 37
— Larionovii Grub. et Vass. 9, 29
— lasiopoda Bge. 18, 58
— latialata P.C.Li 11, 22
— latibracteata Jurtz. 14, 44, 45
— *Lavrenkoi* Ulzij. 71
— Lehmannii Bge. 11, 22
— leptophylla (Pall.) DC. 17, 53
— *leucocephala* Ulbr. 35
— *leucopodia* Ledeb. 70
— *leucotricha* Turcz. 54
— linearibracteata P.C.Li 8, 29
— *longialata* P.C.Li 21
— longibracteata Kar. et Kir. 14, 45
— lutchensis Franch. 7, 29
— *macrobotrys* Bge. 50
— macrocarpa Kar. et Kir. 15, 40
— macrosema Bge. 14, 45
— Martjanovii Kryl. 15, 45
— Meinshausenii C.A. Mey. 11, 35,
36, 37, 38
— melanocalyx Bge. 10, 37
— melanotricha Bge. 7, 29
— merkensis Bge. 8, 28, 30
— — var. ervicarpa (Vved.) Vass. 30
— micrantha Bge. ex Maxim. 17, 53
— microphylla (Pall.) DC. 20, 68
— *microphylla* auct. non DC. 66
— *microsphaera* auct. non Bge. 41
— minutiflora Jurtz. 65
— mixotriche Bge. 16, 53
— mongolica Kom. 18, 59
— monophylla Grub. 16, 54
— *montana* auct. non DC. 37
— muricata (Pall.) DC. 20, 69
— myriophylla (Pall.) DC. 18, 59
— *neimongolica* C.W. Chang et
Y.Z.Zhao 54
— nitens Turcz. 17, 54
— *nivalis* Franch 20
— nutans Bge. 7, 30
— ochrantha Turcz. 17, 60
— ochrocephala Bge. 10, 36, 37
— — var. *longibracteata* P.C.Li 60

— ochroleuca Bge. 10, 38
— oligantha Bge. 19, 60
— *oligantha* auct. non Bge. 57
— oxyphylla (Pall.) DC. 19, 61
— *oxyphylla* auct. non DC. 65
— *pagobia* Bge. 21, 22
— *pamirica* Danguy 27
— parasericopetala P.C.Li 11, 22
— *paratragacanthoides* Vass. 74
— pauciflora Bge. 11, 22
— Pavlovii B. Fedtsch. et Basil. 19, 61
— pellita Bge. 18, 62
— penduliflora Gontsch. 7, 31
— *physocarpa* auct. non Ledeb. 66
— pilosa (L.) DC. 10, 36, 39
— platonychia Bge. 9, 39
— platysema Schrenk. 11, 23
— podoloba Kar. et Kir. 12, 49
— *Politovii* Sumn. 60
— *polyadenia* Freyn 66
— *Poncinsii* Franch. 56
— Potaninii Bge. ex Palib. 17, 54
— proboscidea Bge. 12, 20, 23
— prostrata (Pall.) DC. 19, 62
— Przewalskii Kom. 19, 63
— *psammocharis* Hance 65
— pseudofrigida Saposhn. 17, 55
— *pseudolanuginosa* Jurtz. 58
— *puberula* Boriss. 33, 34
— pulvinata Saposhn. 50
— pulvinoides Vass. 13, 49
— pumila Fisch. ex DC. 18, 63
— *pumila* auct. non Fisch. 65
— pusilla Bge. 7, 31
— racemosa Turcz. 19, 66
— ramossisima Kom. 17, 64
— recognita Bge. 13, 46
— rhinchophysa Schrenk 18, 64
— rhizantha Palib. 16, 55
— *rigida* M. Pop. 57
— *robusta* M. Pop. ex Vass et B.
Fedtsch. 40
— rupifraga Bge. 7, 31
— sacciformis H.C.Fu 20, 71
— *salina* Vass. 34
— Saposhnikovii Kryl. 7, 31
— sarkandensis Vass. 12, 49

Oxytropis saurica Saposhn. 19, 64
— savellanica Bge. ex Boiss. 11, 24
— Schrenkii Trautv. 13, 50
— selengensis Bge. 19, 64
— Semenovii Bge. 15, 40
— sericopetala C.E.C.Fisch. 11, 24
— setifera Kom. 16, 55
— setosa (Pall.) DC. 16, 55
— soongorica (Pall.) DC. 14, 46
— *spinifer* Vass. 74
— squamulosa DC. 12, 70
— *stipulosa* Kom. 25
— Stracheyana Benth. ex Baker 17, 56
— strobilacea Bge. 14, 45, 46
— — var. *chinensis* Bge 44
— — var. *mongolica* Bge. 44
— *strobilacea chinensis* Bge. 44
— *subfalcata* Hance 25
— *Sumnewiczii* Kryl. 64
— sutaica Ulzij. 18, 65
— taldycola Grub. et Vass. 15, 40
— tatarica Camb. ex Bge. 12, 24
— *tatarica* auct. non Camb. ex Bge. 71
— tenuis Palib. 12, 50
— tianschanica Bge. 13, 50
— *tibetica* Bge. 66
— *Tilingii* Bge 29
— tragacanthoides Fisch. 6, 74
— *transversa* Vass. 47
— *trichocalycina* Bge. 72

— trichophysa Bge. 19, 69
— *trichophysa* auct. non. Bge. 67
— trichosphaera Freyn 20, 71
— tschujae Bge. 14, 47
— Ulzijchutagii Sancz. 10, 39
— *uralensis* Ledeb. 42
— *uralensis* auct. non DC. 44
— *uratensis* Franch. 56
— Valerii Vass. 15, 41
— viridiflava Kom. 18, 65
Oxytropis, sect. 9, 20
Oxytropis, subgen. 7, 20, 36
Phaca lanata Pall. 58
— *lapponica* Wahl. 36
— *microphylla* Pall. 68
— *muricata* Pall. 69
— *myriophylla* Pall. 59
— *prostrata* Pall. 62
Phacoxytropis Bge., subgen. 20
Physoxytropis Bge., subgen. 20, 70, 71
Polyadena Bge., sect. 17, 66, 68
Protoxytropis Bge., sect. 20, 22
Ptiloxytropis Bge., subgen. 20, 71
Ramosissima Grub., subsect. 64
Sphaeranthella Gontsch., sect. 15, 40
Traganthoxytropis Vass., subgen. 6, 72
Triticaria Vass., subgen. (Oxytropis) 20, 72
Xerobia Bge., sect. 15, 50

INDEX OF PLANT DISTRIBUTION RANGES

	Map
Oxytropis bungei Kom.	1
Oxytropis chionophylla Schrenk	3
Oxytropis densa Benth. ex Bge.	4
Oxytropis falcata Bge.	2
Oxytropis glacialis Benth ex Bge.	1
Oxytropis imbricata Kom.	3
Oxytropis kansuensis Bge.	1
Oxytropis mongolica Kom.	2
Oxytropis monophylla Grub.	2
Oxytropis pavlovii B. Fedtsch. et Basil.	4
Oxytropis valerii Vass.	3

INDEX OF PLANT DRAWINGS

	Plate	Fig.
Oxytropis coerulea (Pall.) DC.	I	4
Oxytropis deflexa (Pall.) DC.	II	3
Oxytropis dichroantha Schrenk.	III	3
Oxytropis eriocarpa Bge.	II	1
Oxytropis globiflora Bge.	I	3
Oxytropis hirsuta Bge.	III	5
Oxytropis hirta Bge.	IV	1
Oxytropis hystrix Schrenk	I	2
Oxytropis kansuensis Bge.	I	1
Oxytropis lasiopoda Bge.	IV	4
Oxytropis macrocarpa Kar. et Kir.	III	2
Oxytropis meinshausenii C.A. Mey.	III	1
Oxytropis monophylla Grub.	II	2
Oxytropis racemosa Turcz.	IV	2
Oxytropis squamulosa DC.	IV	5
Oxytropis strobilacea Bge.	II	4
Oxytropis sutaica Ulzij.	III	4
Oxytropis trichophysa Bge.	IV	3